Biochemical Research Techniques:
A Practical Introduction

Biochemical Research Techniques:
A Practical Introduction

Edited by
JOHN M. WRIGGLESWORTH

Chelsea College, University of London, UK

A Wiley–Interscience Publication

JOHN WILEY & SONS
Chichester · New York · Brisbane · Toronto · Singapore

Library of Congress Cataloging in Publication Data:
Main entry under title:

Biochemical research techniques.
Includes index.
1. Biological chemistry – Technique.
I. Wrigglesworth, John M.
QP519.7.B55 1983 574.19′2′072 82-21963

ISBN 0 471 10323 3

British Library Cataloguing in Publication Data:

Biochemical research techniques.
1. Biological chemistry – Technique.
I. Title II. Wrigglesworth, John M.
574.1912′028 QP519.7

ISBN 0 471 10323 3

Typeset by Oxford Verbatim Limited
Printed by Page Bros., (Norwich) Limited

List of Contributors

GHEORGHE BENGA — *Department of Cell Biology, Faculty of Medicine, Medical and Pharmaceutical Institute, Cluj-Napoca, Romania*

ALAN H. BITTLES — *Department of Human Biology, Chelsea College, University of London, London SW3 6LX, UK*

ROBERT F. G. BOOTH — *Biology Department, Searle Research & Development, High Wycombe, Bucks. HP12 4HL, UK*

PETER NICHOLLS — *Department of Biological Sciences, Brock University, St. Catharines, Ontario L2S 3A1, Canada*

PETER J. QUINN — *Department of Biochemistry, Chelsea College, University of London, London SW3 6LX, UK*

NORMAN A. STAINES — *Immunology Section, Chelsea College, University of London, London SW3 6LX, UK*

W. PATRICK WILLIAMS — *Department of Biophysics and Bioengineering, Chelsea College, University of London, London SW3 6LX, UK*

JOHN M. WRIGGLESWORTH — *Department of Biochemistry, Chelsea College, University of London, London SW3 6LX, UK*

Contents

Preface

The main purpose of this book is to help research students expand their competence in techniques they might not yet have applied to their research problems. It should also prove helpful to established research workers who wish to become more familiar with techniques outside their speciality. Without a great deal of effort it is often quite difficult for a student to assess whether, in practice, an unfamiliar technique would be useful in helping to solve a particular problem. Research is inevitably concerned with the minutiae of technical detail, and research papers and review articles usually presuppose a background experience of technical practice.

There are many specialist reviews on each of the techniques covered here but most are too detailed to provide a useful practical introduction to the subject. Because of this, each author was asked to present a brief overall view of the technique including any necessary theory but then should concentrate on practical details (sample preparation, signal treatment, etc.) to enable the reader to assess whether the technique would provide useful information in his or her particular line of research. Finally the chapters illustrate the range of use of the technique together with reference to more detiled texts.

Research students in biochemistry are expected to gain some familiarity with a diverse range of experimental techniques. It would be impracticable to attempt to cover all in one volume and the present selection reflects the interests of the editor. Although the topics range over a wide field, from the atomic (spectroscopy) through the molecular (spin labelling and chromatography) to the cellular (monoclonal antibody production and cell culture), they have been chosen to cover areas of common interest to many laboratories in the biological and medical sciences. The topics range from methods in regular use where expert advice on how to extract the best results is often difficult to find, to some of the newer techniques where advice is given on potential applications.

JOHN M. WRIGGLESWORTH
1982

Biochemical Research Techniques
Edited by J. M. Wrigglesworth

1
Absorbance Spectroscopy

PETER NICHOLLS

Department of Biological Sciences, Brock University, St. Catharines, Ontario L2S 3A1, Canada

---

1.1 Introduction

Absorbance spectroscopy with visible and ultraviolet light is the most widely used and least formalized of the spectroscopic techniques available to biochemists. Contemporary instruments have evolved from two types originating during the nineteenth century – the hand spectroscope (or microspectroscope) and the visual colorimeter. The qualitative or semiquantative hand spectroscope was used by Hoppe-Seyler and by Stokes in the 1860s to demonstrate the various functional states of haemoglobin as well as the complexity of plant pigment composition (see Keilin, 1966). The colorimeter, originating as a simple comparator block using white light, was elaborated by the introduction of the monochromator and the photocell. The earliest haemoprotein spectra were produced either by camera plus microspectroscope (Kurt Stern's demonstration of catalase peroxide complexes in 1936, the earliest direct proof of enzyme substrate combination) or by comparative colorimetry using a monochromator (cf. Keilin and Hartree's catalase spectra of 1945 using the Hilger–Nutting spectrocolorimeter). The writer has in fact used the latter instrument to obtain a fairly accurate visible spectrum of reduced cytochrome *c*; its usefulness is limited by tiring of the human eye presented with a series of coloured fields of progressively changing wavelengths whose halves must be manually balanced for brightness. (Keilin reported that on one occasion continued use of the machine brought on the temporary loss of part of the visual field in his eye.) Beckman's combination of photocell and monochromator made the human eye obsolete in spectroscopy although the initial Beckman DU and the Hilger Uvispek instruments still required electrical null balancing against a reference cell at each wavelength (the writer's 1959 thesis must have been one of the last whose spectroscopic and kinetic data were obtained entirely using the null balancing method – a certain dexterity was needed in following reactions at 5 sec intervals, but the optical quality of stable scanned spectra, with 'smoothing' feedback via the operator, has only recently been exceeded by photomultiplier-dependent automatic instruments).

The increasing sophistication of the spectrophotometer, with the introduction of the more sensitive photomultiplier in place of the photocell and automatic dual beam operation and baseline correction, led to the disappearance of the microspectroscope and camera as research tools, despite their potential advantage in terms of the rate of data acquisition across a complete spectral range. The renewed need for rapid data acquisition and the ability to store much data in computer memories rather than on film, has led to the recent reintroduction of automatic instruments based on the microspectroscope principle of illuminating the sample with white light and analysing the absorption changes at numerous wavelengths simultaneously using an array of photodiodes and a storage memory in place of a single photomultiplier and chart recorder. Examples of spectra obtained with monochromator-based scanning spectrophotometers and with diode array instruments are included in the following pages.

1.2 Principles and Techniques

The theory of light absorption by biomolecules is described in Chapter 2 on fluorescence (Williams, 1983). Absorption spectrophotometers differ from fluorimeters in monitoring changes in transmitted light rather than the production of fluorescent light, although some basic principles of construction remain the same (see Fig. 1–4 below with Figs. 4–7 in Chapter 2).

Fig. 1 shows schematically the principal components of standard spectrophotometers. In Fig. 1A we see the optical arrangement of lamp (a), monochromator

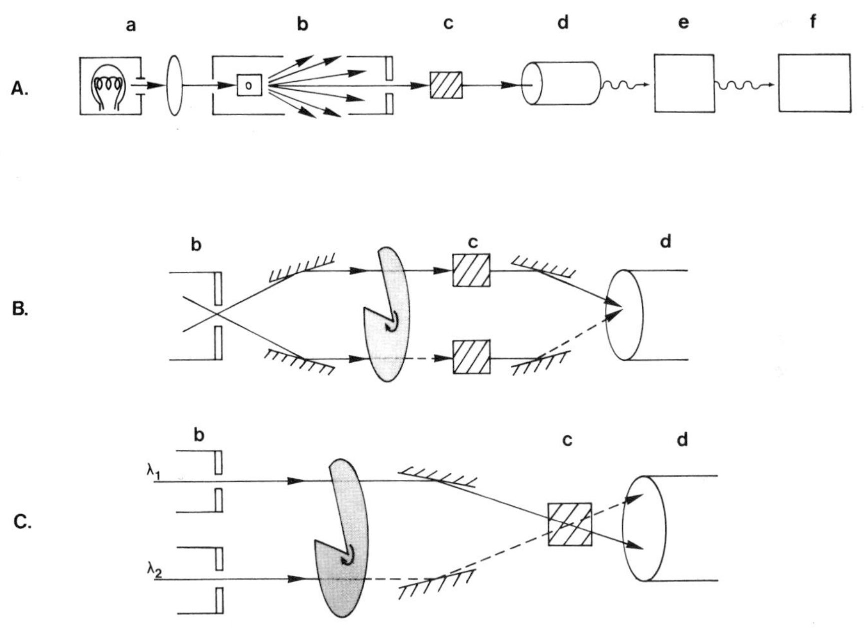

Fig. 1 Standard spectrophotometer systems: schematic diagrams. A. Single beam spectrophotometer: a, light source; b, monochromator; c, sample compartment; d, detector; e, amplifier; f, recorder. B. Split beam spectrophotometer (alternative to b–d in Fig. 1A): b, monochromator generating two beams, which pass a 'chopper' that alternates between the two beams; c, sample and reference cell compartment, arranged so that one beam passes each cell; and d, detector system responding to the difference in intensity of the two beams. C. Dual wavelength spectrophotometer (alternatives to b–d in Figs 1A and 1B): b, double monochromator generating two beams of different wavelengths, 'chopped' as in 1B; c, sample cell arranged so that the two beams pass alternately; and d, 'end-on' detector responding to the difference in intensity of the two beams

(b), cell/sample compartment (c), detector (d), amplifier (e) and recorder (f). In the conventional single beam instrument, standardized around 1950, the transmitted light for the reference sample is arbitrarily set at 100% for each wavelength and the light transmitted by the sample then measured at the same wavelength. The split beam instrument (standard since about 1960) passes light alternately through sample and reference sample (Fig. 1B) and the detector/amplifier system is used to measure the differential transmittance of the two signals. Automatic scanning with such an instrument requires mechanical or electronic baseline flattening over the wavelength range scanned. The dual wavelength instrument (Fig. 1C) was pioneered by Britton Chance and has been commercially available since about 1965. In this type of spectrophotometer light from two monochromators passes alternately through a single sample and the differential absorbance is measured as in Fig. 1B.

The alternative layout adopted for photodiode array instruments is illustrated in Fig. 2. Here the sample compartment (c) precedes the grating system (b) that replaces the monochromator; all the dispersed light is detected by the array (d) and the signal collection device (e) has to be that much more elaborate. A complete (400 nm) spectrum can however be collected within milliseconds.

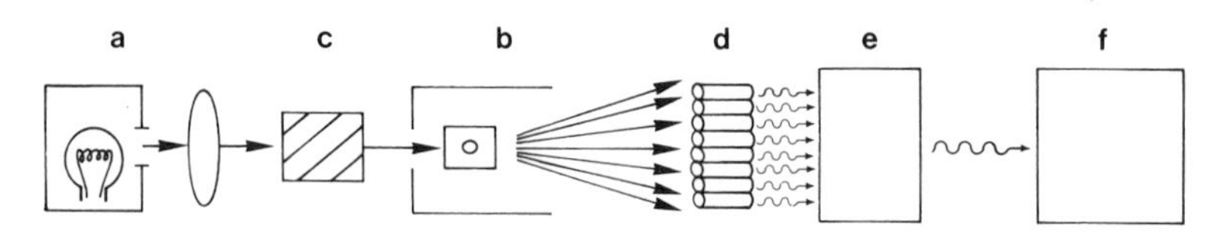

Fig. 2 Diode array spectrophotometer system: schematic diagram: a, light source; b, grating; c, sample compartment; d, diode detector array; e, amplifier and multiple memory system; f, recorder system (storage oscilloscope or printer-plotter)

Faster data collection from standard instruments can only be achieved by measurement at single wavelengths or wavelength pairs. The problem is then the rapid initiation of a chemical process within the sample. Various kinds of rapid mixing devices have been invented, starting from the original continuous flow apparatus of Hartridge and Roughton (1923). Fig. 3 shows the typical layout of the most common arrangement, the stopped flow machine. Non-chemical ways of starting reactions include temperature and electrical jump devices as well as flash techniques for photosensitive reactions, illustrated in Fig. 4. A comprehensive review of fast enzyme reactions and the technique required to study them has been written by Hiromi (1979).

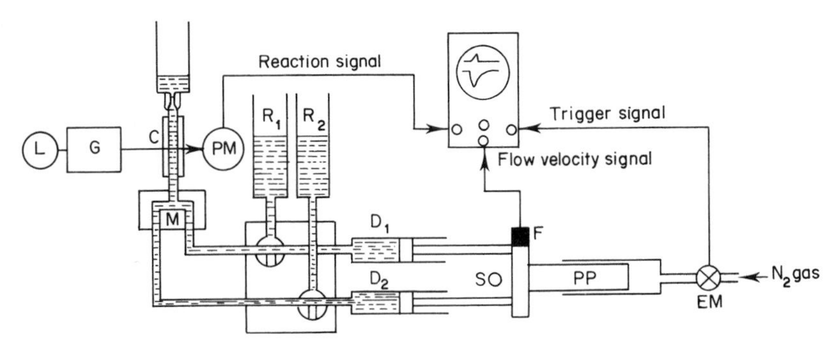

Fig. 3 Stopped-flow spectrophotometer: L, light source; G, grating monochromator; C, sample cell; M, mixing chamber; PM, detector; D_1, D_2, driving syringes; F, flow velocity detector; EM, electromagnetic valve; PP, gas-driven piston; S, stopping pin.*
*Front-stopping type; other models (e.g. the Durrum–Gibson) use a back-stopping device after the collecting syringe (reproduced with permission from Hiromi, 1979)

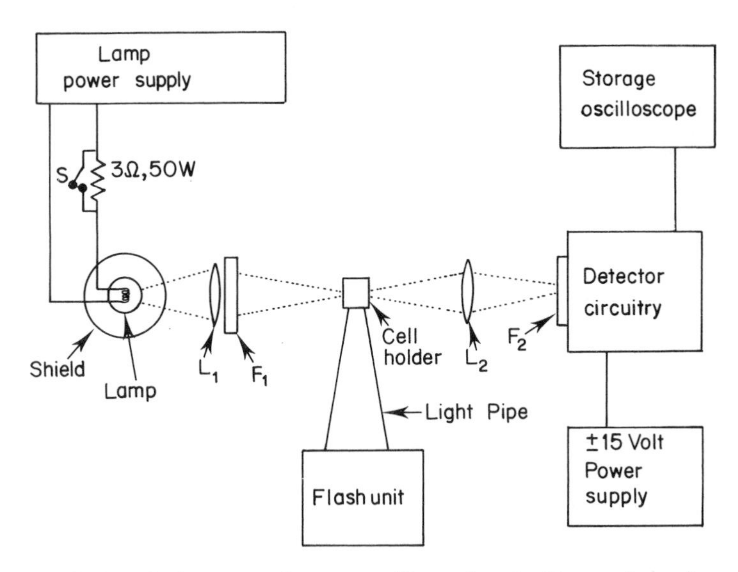

Fig. 4 Flash photolysis spectrophotometer (Reproduced with permission from Sawicki and Morris, 1981). Lamp, light source for observation beam; L_1, L_2 focusing lenses; F_1, F_2, filters to secure appropriate measuring light and to eliminate flash light interference; S, stopping pin. Flash unit – photographic flash system with time constant of 0.2 to 1 msec

1.3 Equipment and Components

Fig. 1 indicates the five essential components of any spectrophotometer. Some comments may be made upon each.

1.3.1 Lamps

All instruments use at least two lamp systems, one for the ultraviolet region (180–350 nm) and one for the visible and near-IR regions (320–2000 nm). The ultraviolet lamps now used are deuterium discharge lamps (originally hydrogen discharge lamps were employed). These have a near-continuous spectrum in the UV region and show emission lines in the visible region which may be used for calibration purposes. Tungsten filament bulbs are commonly used as visible lamps, with appropriate voltage stabilization: they are often the least specialized part of the instrument and care must be taken when replacing lamps to check filament position and alignment. One standard line of commonly used bulbs, for example, was recently changed from a vertical to a horizontal filament (retaining its original stock number!), thereby preventing proper alignment with the entrance slit in the monochromator. The quartz iodine lamp is an improvement over the simple tungsten lamp; spectral gaps render the more powerful xenon lamps less useful spectrophotometrically. All such lamp systems benefit from being connected to their power supply via a voltage stabilizer. Not all transformer boxes supplied commercially act as effective stabilizers, and in doubtful cases an extra such device in line is always recommended.

1.3.2 Monochromators

White light passed through the entrance slit of the monochromator is dispersed into its monochromatic components by a prism (classical) or a grating (contemporary). Gratings are optically more efficient than prisms, and give a more even dispersion of the light, but give secondary light at wavelengths which are an integral multiple of the wavelength desired. For this and for other reasons, filters are employed which restrict the transmitted light to specific spectral regions. These may have the disadvantage of reducing the intensity of the desired light and of introducing anomalous baseline shifts into continuously scanned spectra. Commercial instruments often have such filters automatically inserted at points of key biochemical interest (575 nm, the peak of oxyhaemoglobins, and 600 nm, close to the peak of reduced cytochrome *c* oxidase, for example).

Stray light (usually essentially white) is emitted by most monochromator systems. To measure a specific absorbance of 4.0 A, stray light has to be at the 0.001% level. Few instruments achieve this. Usually it is simplest to use a more dilute sample or a shorter light path.

The exit slit of most monochromators can be adjusted to a given band width. Colorimeters characteristically use band widths of 6–20 nm, while true spectrophotometers use band widths of 0.1–3 nm. 1 or 2 nm would be usual, few biological absorption bands being narrow enough for appreciable differences to be seen on going to narrower band widths. Discrimination between close peaks depends upon the peak widths as well as the spectrophotometer parameters.

1.3.3 Sample Compartment

Depending upon the nature of the spectrophotometer system employed, the sample space will contain one cuvette with a single transmitted beam (Fig. 1A), two cuvettes with a beam passing alternately through each (Fig. 1B), or one cuvette with two beams (of different wavelength) passing alternately through it (Fig. 1C). In all cases, the usual cuvette is of a square cross-section (1 cm × 1 cm) and total volume around 3 cm^3. Smaller cuvettes can be made by reducing the light path (0.2 or 0.1 cm) *or* by reducing the size of the measuring beam and that of the sample using cells of smaller height or width. Masks are usually inserted into the light beam. In dual cuvette systems, where the beam shape passing the two cuvettes is rarely identical, this will alter the baseline position. The smallest useful volume that can be measured in a semimicro cell system is about 0.1 ml. Specialized microspectrophotometers capable of, for example, measuring the spectrum of a single red cell, have a different, precision beam geometry and are not commonly used. Fig. 18 illustrates the results that can be obtained with such an instrument capable of monitoring the spectrum of all or part of a single cell (see also examples and discussion in Nicholls and Elliott, 1974). Commercial instruments of this type have been produced, notably by Shimadzu.

Temperature control of the sample is always necessary in kinetic studies and often in scanning spectra. Jacketed cells are available, but in the usual temperature range (15–40 °C) a temperature-controlled cell holder is adequate. The temperature in the cuvette should be checked at the end of the experiment to compare with the set temperature of the circulating water bath. Condensation from the atmosphere usually limits observations with conventional cuvettes at temperatures below ambient.

1.3.4 Detectors

The photomultiplier (with power supply required) has effectively replaced the old photo tube. Most photomultiplier tubes appropriate for use in the ultraviolet region have a poor red end response and for work at wavelengths greater than 750 nm (where a number of important biomolecules absorb light) a specially selected tube is required or a lead sulphide cell capable of response to near-infra red light (out to 2500 nm or beyond). Most commercial instruments are equipped with 'side-on' window photomultipliers; for work with turbid solutions an instrument with an 'end-on' window photomultiplier is to be preferred. The window should be flat (many PM tubes with good responses but designed for other purposes have curved entry windows). The dynode chain should be compatible with the amplifier system. Although photomultiplier tubes have greater sensitivity and usually a more rapid response than phototubes, care must be taken in exposing them to light, especially when the high voltage is connected. Light 'shock' can cause a decrease in sensitivity with rather slow recovery.

1.3.5 Recorders and Amplifiers

The signal from the photocell or photomultiplier is accepted and amplified for storage in a memory or by a mechanical recorder. All diode array instruments demand electronic memory storage prior to mechanical readout. Some mechanical scanning systems (e.g. the Beckman DU-7) also provide for storage but others (e.g. the Cary-Varian series), although microprocessor-controlled, permit such storage only with an accessory computer. Recorder limitations often slow the data acquisition rate; absorption bands such as the α peak of reduced cytochrome *c* may require an excursion to maximum absorbance within 5 nm, necessitating a scan speed below half this value per second for recorders with commonly available time constants of a second or so for full scale excursion. In addition electronic smoothing often introduces time constants of the order of tenths of seconds or even seconds to reduce the noise level of commercial instruments to values below 10^{-3} A (top quality machines may be expected to show noise levels around 2×10^{-4} A in the visible region). Highly accurate spectra can be obtained with electronic smoothing at very low scan rates (0.01 nm/sec) offered for analysis of organic compounds with sharp UV spectra. Such methodology is less useful for aqueous solution chemistry of light- and thermosensitive biological species. It may thus be preferable, even with mechanical scanning instruments, to use electronic memory and/or oscilloscope readout. Final 'hard copy' is then traced on a printer plotter after the scan itself is over. Such devices will probably soon render the direct recorder obsolete. Whether the complexities of microprocessor 'programme' control really have any advantages over the traditional multiple switching systems is however much more doubtful.

1.4 Methodology of Measurement

1.4.1 Transparent Solutions

For any given quantity of light-absorbing material in the sample beam, the percentage reduction in the amount of light transmitted will always be the same. An increase in concentration, unaccompanied by any chemical change, will produce the same effect as a proportionate increase in depth of absorbing solution. Therefore if the transmitted light, I, is a given fraction of the incident light, I_0, a doubling of concentration or a doubling of sample depth will change I/I_0 to $(I/I_0)^2$. The logarithmic function, $\log I_0/I$, will thus change to $\log (I_0/I)^2$ or $2 \log (I_0/I)$. This function therefore is proportional to concentration and to sample depth according to Eq. 1:

$$\log (I_0/I) = A = \varepsilon Lc \quad (1)$$

where A = absorbance (a dimensionless number), L = sample depth (cm), c = concentration (moles litre^{-1} or M), and ε is a constant, with dimensions M^{-1} cm^{-1}, termed the extinction coefficient, characteristic of the substance and solvent involved for the temperature and wavelength employed.

Fig. 5 illustrates the behaviour of a solution obeying Eq. 1, the Beer–Lambert law, such as an aqueous solution of a non-aggregating dye. An exponential decline in % T (percentage transmission) is equivalent to a linear increase in absorbance, A, over a twenty-fold concentration range. Deviations from this relationship may occur for instrumental or for chemical reasons. The stray light effect will cause the absorbance to flatten out to a fixed value at high concentrations as the % T falls to a fixed value. Thus a stray light of <0.1% at 340 nm (Beckman DU-7 brochure) implies absorbances of up to 3.0 A can be measured before apparent deviations from Eq. 1 occur while a value <0.0008% (Varian 2200/2300 brochure) up to 600 nm implies that absorbances of up to 5.1 A are in principle measurable, provided that the photomultiplier system is sufficiently sensitive, without deviation from Eq. 1. Optical deviations from the Beer–Lambert law will be the same whether L or c is the factor varied. Chemical deviations can be identified when Eq. 1 is obeyed for variations in L (cuvette path lengths from 0.1 to 100 mm can be used) but not for variations in c (sample concentration). An example is shown in Fig. 6, in which the apparent extinction coefficient at 395 nm for coproferrihaem declines from 155,000 M^{-1} cm^{-1} at 40 mM (corresponding to A = 0.006 in a one cm cell) to about 60,000 M^{-1} cm^{-1} at 0.32 mM (corresponding to A = 1.9 in a one mm cell). Such deviations provide evidence for intermolecular interactions (cf. discussion of Fig. 13B below), in this case dimer formation.

As there are only two species in the system, the sum of which remains constant, the apparent extinction coefficient at any point at which the extinction coefficients are equal will be independent of concentration (i.e. at these wavelengths the Beer–Lambert law will be obeyed, as at 370 nm or 405 nm in Fig. 6). Such a point is termed an isosbestic point. Similar points are seen in any set of spectra in which one species is transformed into another, as illustrated in Fig. 7. Fig. 7A illustrates the transformation of a mixture of oxidized cytochrome c and a into their corresponding reduced forms, plotted as absolute spectra in the visible region. Isosbestic points on each side of the α and β peaks

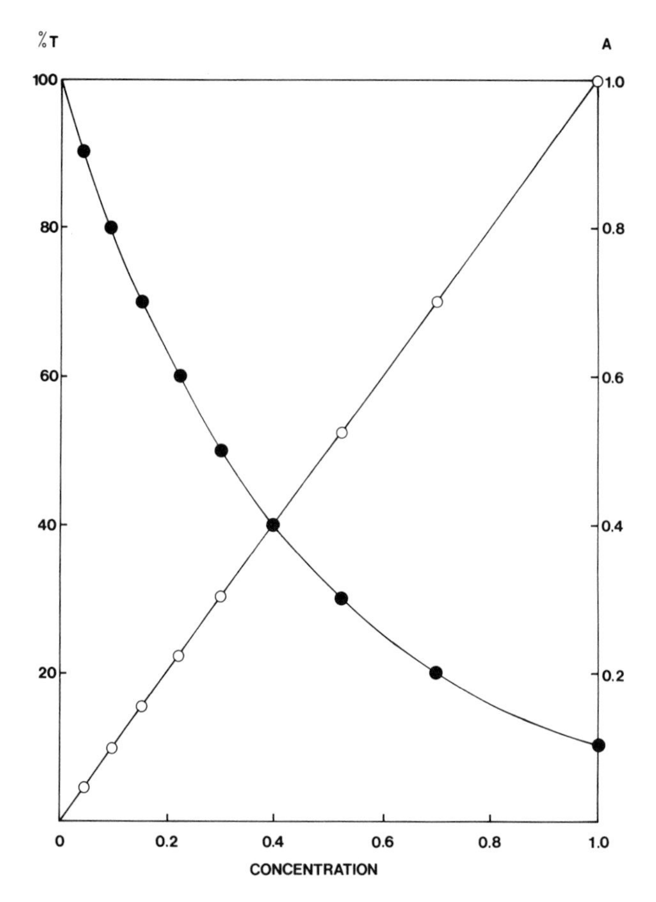

Fig. 5 The Beer–Lambert law exemplified. Absorbance and per cent transmission at a characteristic wavelength of a series of concentrations of coloured substance in aqueous solution. ●, % T (transmission) against concentration; O, *A* (absorbance, log I_o/I) against concentration

of cytochrome *c* indicate that it is titrating as a single species with two forms (ferri- and ferro-) and that *either* cytochrome *a* is without significant absorbance in this region *or* that it is titrating with effectively the same redox potential as the cytochrome *c*. A reciprocal argument may be used for the isosbestic point at 620 nm; however the failure of the 580–590 nm region to show an isosbestic point indicates that a third component or form is involved in this region.

Isosbestic points and deviations therefrom are more readily seen in difference spectra where the spectrum of the starting material is subtracted from all subsequent spectra. Fig. 7B shows the spectra obtained during a slow ligand

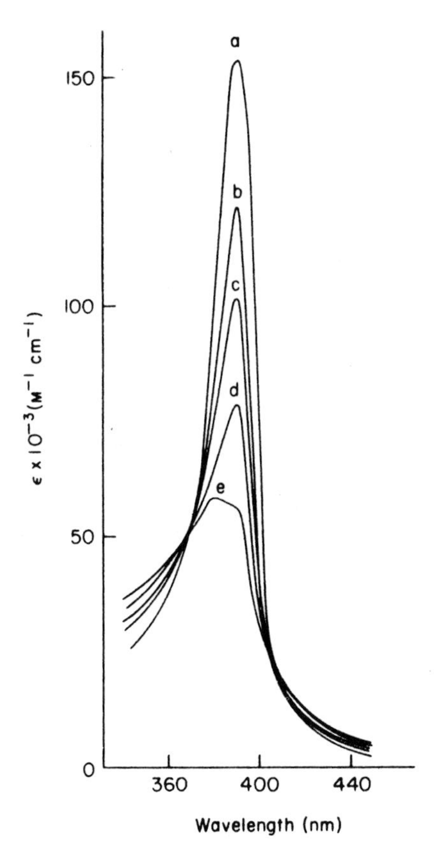

Fig. 6 Deviation from the Beer–Lambert law. Spectra of coprohaematin in phosphate buffer at pH 7 (Soret region) a, 40 nM; b, 6.4 μM; c, 28 μM; d, 0.1 mM; e, 0.32 mM coprohaematin. Reproduced with permission from Brown (1980), *An Introduction to Spectroscopy for Biochemists*. Copyright: Academic Press Inc. (London) Ltd

binding process, in this case the reaction of cyanide with cytochrome a_3 in cytochrome *c* oxidase. Isosbestic points at 337, 423, 474, 522, 609 and above 700 nm show that a single transition is involved. In addition to the requirement for stable isosbestic points (which might be seen for a three-component system in which the third component coincidentally had an absorption equal to that of the other species at certain wavelengths) it may be noted that the difference spectra must all be similar to the Euclidean sense – multiplying any spectrum by a certain factor must give rise to any other.

Fig. 7C shows the deviations from isosbestic behaviour that occur in a ligand binding process involving side reactions. The sequential addition of aliquots of sulphide to a solution of metmyoglobin gives rise to the complex sulphide–metmyoglobin and in the blue-green region isosbestic points a and b can be

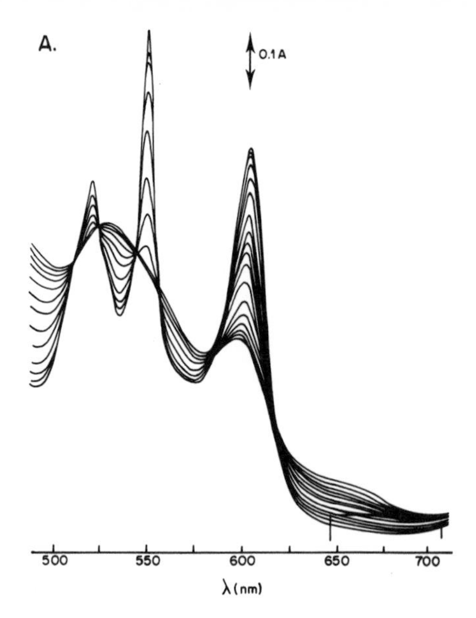
A.
0.1 A
500
550
600
650
700
λ (nm)

B.

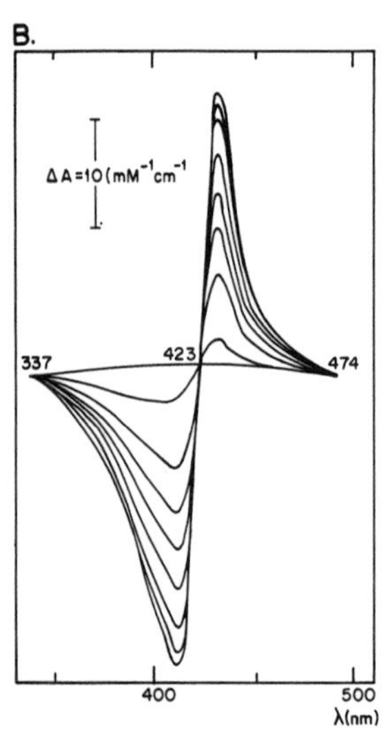
ΔA = 10 (mM⁻¹cm⁻¹
337
423
474
400
500
λ(nm)

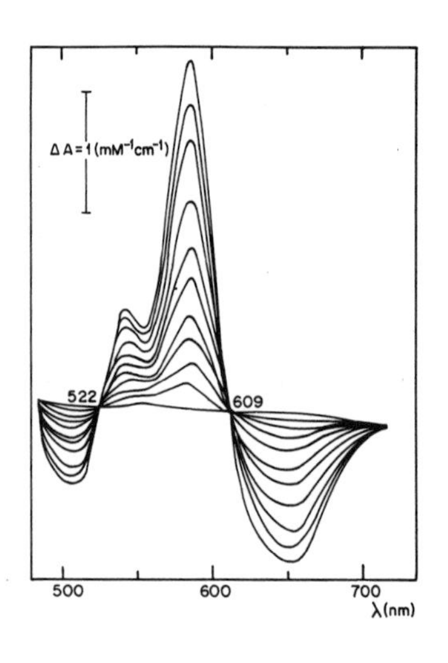
ΔA = 1 (mM⁻¹cm⁻¹)
522
609
500
600
700
λ(nm)

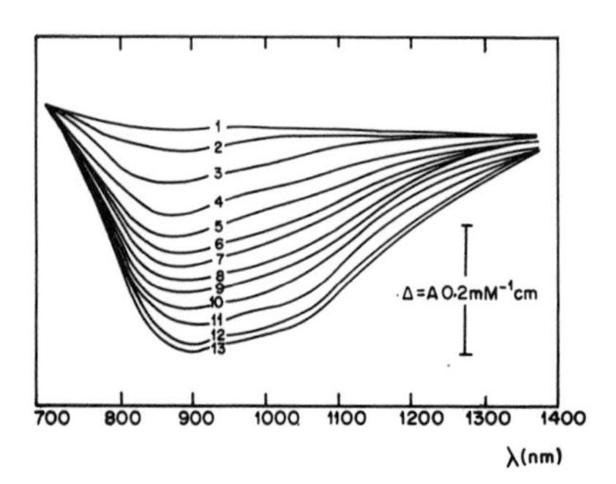
1
2
3
4
5
6
7
8
9
10
11
12
13
Δ = A 0.2 mM⁻¹cm
700
800
900
1000
1100
1200
1300
1400
λ(nm)

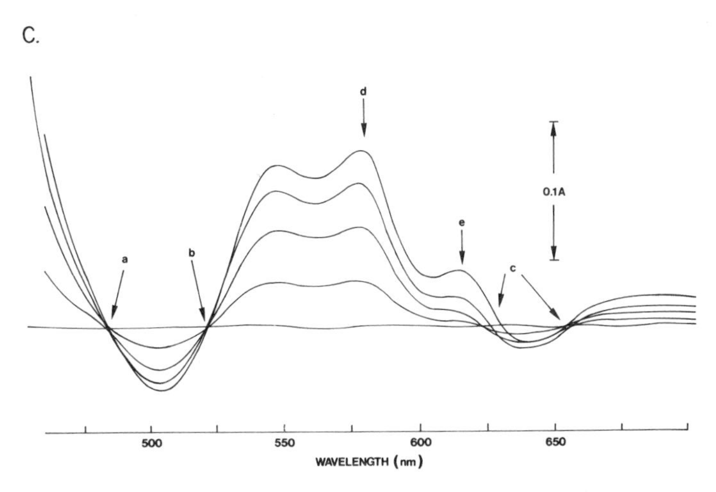

Fig. 7 Patterns of isosbestic points: examples from absolute and difference spectra. A. Spectra of mixtures of cytochrome *c* (peak at 550 nm) and cytochrome aa_3 (peak at 605 nm) during progressive reduction (coulometry). 16 μM cyt. aa_3 plus 21.5 μM cyt. *c* in phosphate buffer with a small amount of detergent and redox mediators, under nitrogen gas. Reprinted with permission from Anderson *et al.* (1976). Copyright 1976 American Chemical Society. B. Difference spectra of cytochrome aa_3 obtained at different times after addition of cyanide (2 mM). 7.5 to 140 μM cyt. aa_3 from 2 minutes to 4 hours after mixing; oxidized cyanide-free enzyme in reference cuvette (reproduced with permission from van Buuren *et al.*, 1972). C. Difference spectra of metmyoglobin obtained after the addition of successive aliquots of sulphide (30 μM sperm whale metmyoglobin plus 0.1 to 1 mM Na_2S in phosphate buffer). a, b indicate isosbestic points; c, d, e indicate a region lacking isosbestic behaviour and demonstrating secondary reactions (Nicholls, 1982)

identified. At the red end of the spectrum, however, isosbesticity is not preserved (region c) due to the secondary formation of the irreversible sulphmyoglobin derivative (peak at e). The peak at d also illustrates the failure of the titration curves to preserve Euclidean similarity, increasing sharpness at increasing sulphide levels indicating reduction of the metmyoglobin to myoglobin (and then oxymyoglobin) instead of simple binding to the ferric form.

When a system contains numerous species, each showing a characteristic spectrum (absolute spectra) or capable of existing in distinct forms (difference spectra), sets of simultaneous equations may be derived for analysing any given preparation for the components it contains: see Williams (1964) for this type of analysis in the case of mixtures of cytochromes. Such equations may now be included in computer programmes for multiwavelength analysis of spectra by instruments (e.g. the Beckman DU-7 and the Varian 2200/2300) that can be controlled by microprocessor. Visual evaluation of all such systems is essential, however, to avoid generating and reporting eccentric results consequent upon light scattering changes or other unexpected anomalies in certain samples.

1.4.2 Turbid Solutions

Many solutions or suspensions of biological interest scatter light as well as absorb it. Light scattering also diminishes the amount of light reaching the photomultiplier. Three important types of light scattering may occur (see also Chapter 2):

(1) Tyndall scattering (Tyndall, 1863);
(2) Rayleigh scattering (Rayleigh, 1899);
(3) Stokes–Raman scattering (see Chapter 2).

Tyndall scattering is that which occurs when the scattering particles are $>\lambda/4$ where λ is the wavelength of the scattered light, while Rayleigh scattering occurs when the diameter of the scattering particles is less than $\lambda/4$. Tyndall scattering is almost wavelength independent, but Rayleigh scattering is dependent on the inverse fourth power of the wavelength, as seen in Fig. 8. This figure compares the scattering by milk (relatively large fat globules) with that by a suspension of liposomes (small sonicated phospholipid vesicles) and by a strong solution of serum albumin. A theoretical curve for Rayleigh scattering is indicated by the dashed line. Both liposomal and albumin systems approach the Rayleigh case, although the sample of albumin used has a high residual (Tyndall) behaviour; milk is an almost pure Tyndall system (i.e. acts as a neutral filter) with a very small Rayleigh component. A plot of log A *vs* log λ can be used to determine the exponent n and the constant B in Eq. 2, the Rayleigh 'coefficient' (theoretically 4) and the Tyndall component in any scattering system:

$$A = B + C/\lambda^n \tag{2}$$

where A is apparent absorbance, λ is wavelength and C is a concentration-dependent term.

Measurement of true absorbance in a scattering system requires placement of the sample close to the photomultiplier, so that minimum losses of the scattered light occur. A cuvette with white side walls can also reduce scattering losses. Fig. 9 shows the spectra obtained for a suspension of red blood cells with an instrument possessing both a distant and a close position for the cuvette relative to the photomultiplier. In the distant position (a) the oxyhaemoglobin spectrum is superimposed upon a large Tyndall scattering component; when the red cells are placed close to the photomultiplier window (b), the scattering is no longer seen (the spectrophotometer is monitoring the scattered light as well) and the measured absorbances in the visible (520–600 nm) region are close to those seen in a haemolysate of the same concentration. However, it should be noted that abolition of the scattered light does not restore the Soret (414 nm) peak to its expected height. This reduction in the apparent Soret extinction coefficient of haemoglobin inside red cells was once thought to

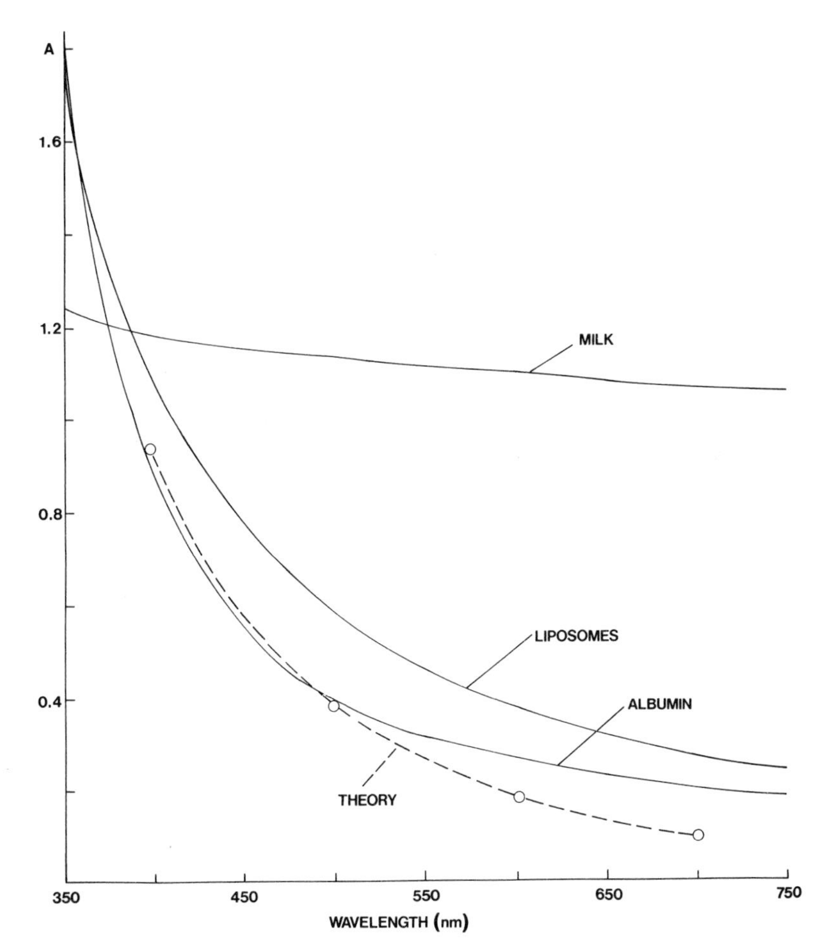

Fig. 8 Three light-scattering curves compared with the predicted Rayleigh behaviour: diluted milk (triglyceride globules), 'Tyndall' (wavelength-independent) scattering; liposomes (phospholipid vesicles, approx. 14 mg ml^{-1} phospholipid mixture in 0.2 M phosphate buffer); protein (20% serum albumin plus 3% H_2O_2 to destroy small amounts of pigment); O---O a theoretical curve for pure 'Rayleigh' (wavelength-dependent) scattering. Pye Unicam SP800 spectrophotometer: 'distant' cuvette position (Nicholls, 1982)

reflect a true difference in chemistry between haemoglobin *in vitro* and *in vivo*. In fact it is an optical artefact consequent upon the packaging of haemoglobin in concentrated 'samples' within the erythrocytes. Internal reflections within the cuvette permit a monochromatic 'stray light' effect, with part of the 414 nm light escaping absorption. An extinction of 0.68 *A* (20% T) instead of 1.7 *A* (2% T) indicates 18% of the incident light being passed forward through the sample without 'seeing' the highly absorbing red cells.

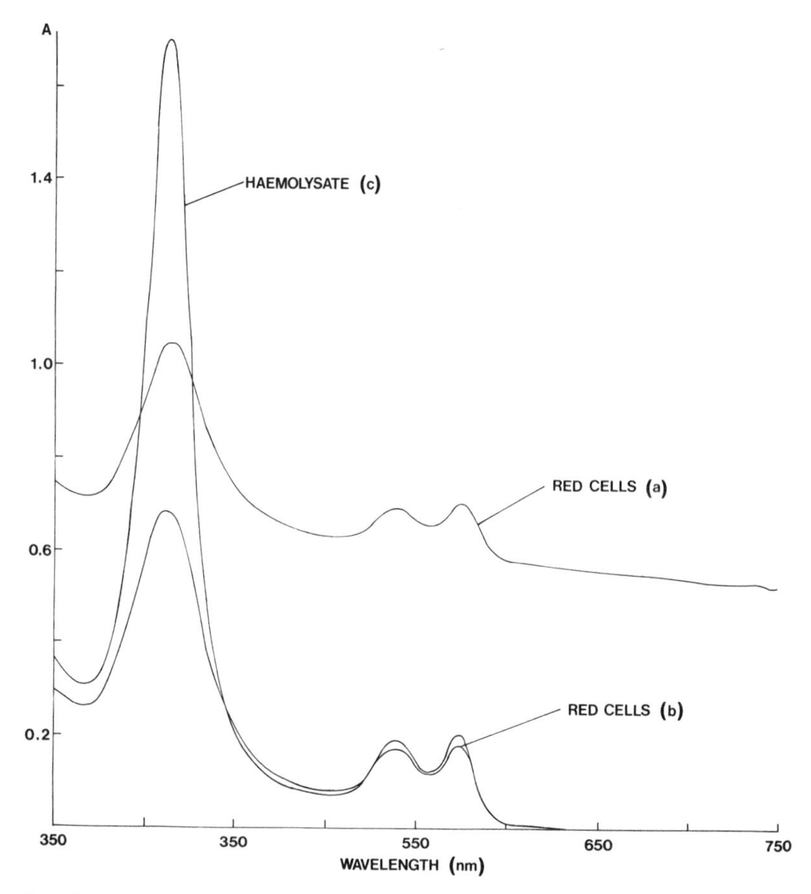

Fig. 9 Light-scattering effects in a pigment-containing system: diluted red cells compared with haemolysate. Red cells (a): human erythrocytes diluted in saline, 'distant' cuvette position; red cells (b): as in (a) but in 'close' cuvette position (near to PM tube); haemolysate (c) (cells as in (a) but with addition of a few mg sodium deoxycholate): 'distant' cuvette position. Note 'increase' in Soret (414 nm) band. Pye Unicam SP800 scanning spectrophotometer (Nicholls, 1982)

The converse effect can sometimes occur with a dissolved pigment in the presence of colourless scattering particles. The latter now increase the effective path length through the sample cuvette and therefore increase the measured extinction coefficient compared with that in transparent solutions (cf. the discussion of low temperature spectra and Fig. 12 below).

With smaller particles showing a large Rayleigh component in their scattering, the use of difference spectroscopy to eliminate the scattering problems becomes essential. Fig. 10A shows the absolute spectra of reduced and oxidized submitochondrial particles obtained with a diode array instrument. Fig.

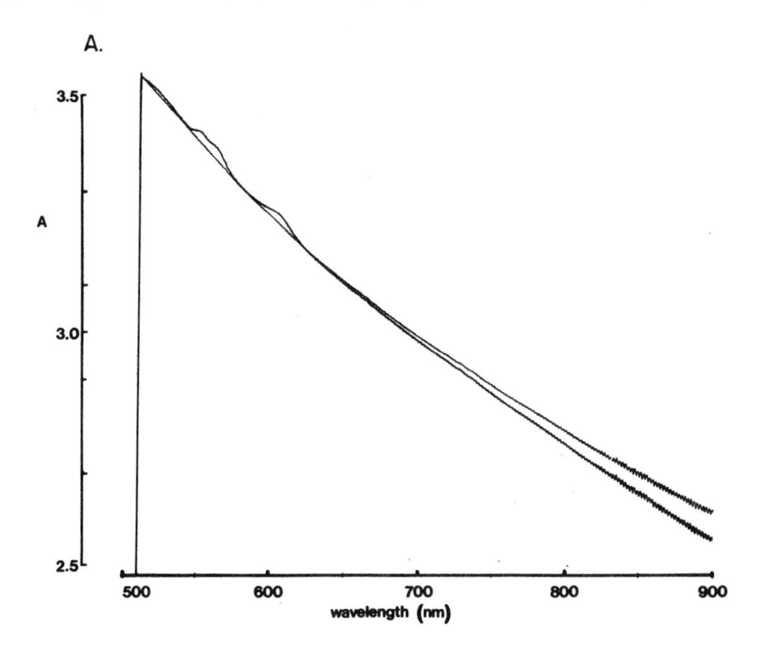

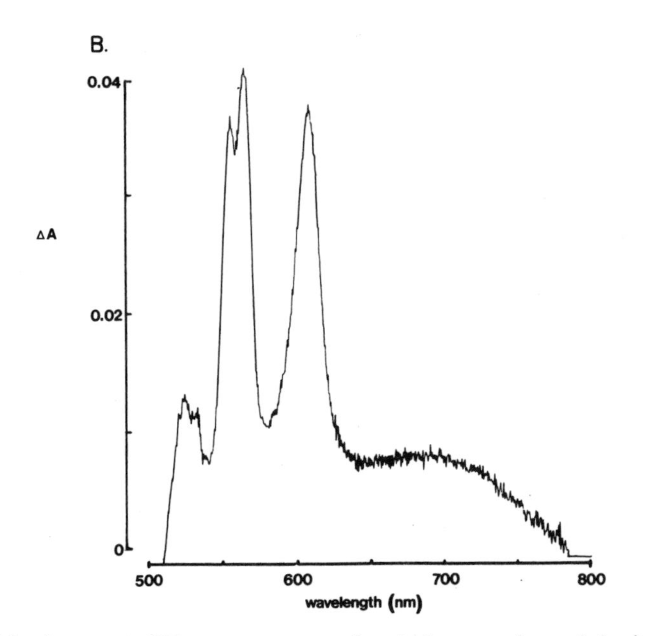

Fig. 10 Absolute and difference spectra of turbid suspensions. Submitochondrial particles (Keilin–Hartree beef heart type) in phosphate buffer treated with $K_3Fe(CN)_6$ (oxidized) or $Na_2S_2O_4$ (reduced). A, absolute spectra of oxidized and reduced samples from 480 to 900 nm; B, difference spectrum (reduced *minus* oxidized) of samples as in A. E.G. & G. (Princeton) diode array scanning spectrophotometer (Nicholls, 1982) (courtesy E.G. & G. Instruments)

10B shows the difference spectrum (reduced *minus* oxidized) obtained from these data as computed by the same machine.

When inhomogenous scattering is a problem, it is often better to 'scramble' the transmitted light before it falls on the photomultiplier, using either a simple translucent 'Shibata' plate or more complex beam scrambling systems (see the Aminco DW-2C instrument). A simple trick that sometimes flattens baselines satisfactorily and puzzles the naive student is to turn the reference and sample cuvettes 90° in the beam, so that the incident and transmitted light pass through the translucent faces of the cuvette, rather than the optical faces.

Under certain conditions both Tyndall and Rayleigh scattering can be used to measure the concentration of scattering particles (B and C in Eq. 2 are both concentration dependent). For any given scattering system a standard curve must be constructed plotting *A vs* concentration for a given wavelength. As with fluorometry (see Chapter 2) the scattered light can also be monitored as it emerges at 45° or 90° to the incident beam. This method (that of 'nephelometry') though more sensitive than the direct measurement of apparent absorbance, is not much used with modern spectrophotometers.

1.4.3 Derivative Spectra

For instrumental and analytical reasons some spectroscopy (e.g. electron paramagnetic resonance) is usually presented in terms of derivative rather than absolute spectra. Multiple components can be detected by analysing absorbance curves into their first and higher derivatives, as shown in Fig. 11. Signal to noise ratio diminishes as higher derivative curves are produced, but this method has been used to disentangle the complexities of multiple cytochrome species in prokaryotes and mitochondria (Shipp, 1971). Second and fourth derivative spectra have similar shapes to the original spectrum (sharpened and containing 'wings'); instrumental signal averaging of some kind is needed if fourth derivative analysis is to succeed (see Shipp, 1971, and references therein).

1.4.4 Low Temperature Spectroscopy

Both optical and chemical properties of absorbing species are modified by temperature change. These changes include: effects upon (1) apparent optical path length, (2) intensity of absorption bands, (3) band widths of absorption bands, (4) position of absorption bands and (5) equilibria and reaction rates.

Optical path length can be increased upon freezing a sample if, instead of a glass, a semitransparent microcrystalline mass is obtained ('devitrification'). The result is a considerable intensification of the absorption bands of the frozen sample (Keilin and Hartree, 1949) superimposed upon the changes due to temperature itself. These latter may include real increases in maximal extinc-

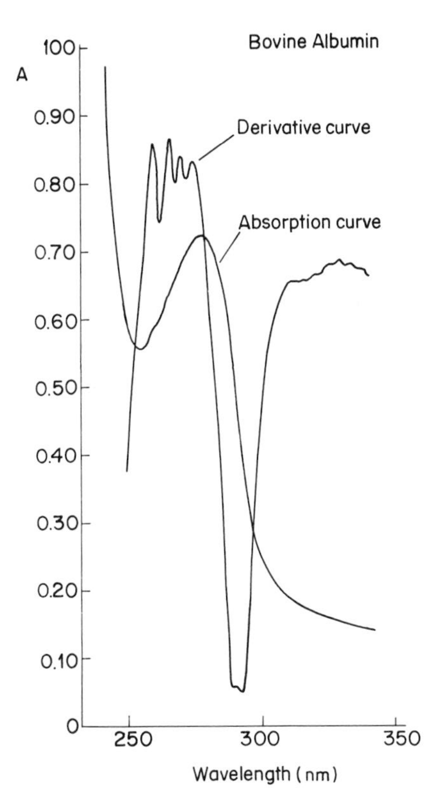

Fig. 11 Derivative spectroscopy: analysis of bovine serum albumin spectrum. Bovine albumin solution in milk. 'Absorption curve': absolute spectrum (see Fig. 16B). 'Derivative curve': first derivative spectrum using single cell and measuring $A(\lambda_1) - A(\lambda_2)$ for fixed small value ($\Delta\lambda$) of ($\lambda_1 - \lambda_2$). Perkin-Elmer 356 dual wavelength instrument (courtesy Perkin-Elmer Co.)

tion coefficient accompanied by decreasing band width ('sharpening', Keilin, 1966) associated with reduction in the gaps between vibrational energy levels at low temperatures. This phenomenon permitted the initial discrimination between cytochromes *c* (α-maximum at 550 nm) and c_1 (α-maximum at 553 nm) in mitochondrial preparations (Keilin and Hartree, 1955). Fig. 12 shows the effect of low temperature on the spectrum of reduced cytochrome *c*. Sharpening of the absorption bands not only permits discrimination between *c* and c_1 but also the analysis of fine structure in β and α bands of individual cytochromes *c* themselves. For example, the α' peak at 547 nm is now distinguished from the main α peak and its position and intensity differ from one cytochrome *c* to the next (it is much reduced, for example, in yeast and insect cytochromes *c* which lack certain amino acid residues always found in the mammalian forms). Fig. 12 also illustrates the visual effect of low temperature on the absorption band

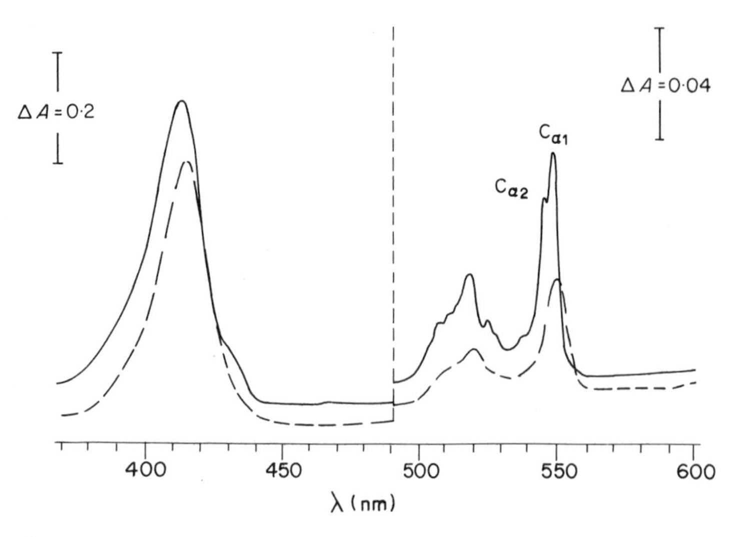

Fig. 12 Low temperature spectroscopy: analysis of cytochrome *c* spectrum. Comparison of room temperature (– –) and liquid nitrogen (—) spectra of solutions of reduced cytochrome *c*: 40 μM cyt. *c* reduced with $Na_2S_2O_4$ in 50% glycerol, 0.05 M phosphate buffer pH 7.4; 1 mm path length cuvette 84 K (reproduced with permission from Estabrook, 1956)

position, a shift towards the blue (indicating an increase in the energy requirement for the electronic transition involved).

More recent studies have taken advantage of chemical effects on spectra at low temperatures. 'Electronic' equilibria, such as those between high and low spin forms of ferric haem proteins, are very temperature sensitive. Alkaline methaemoglobin, mixed spin at room temperature, and azide catalase, high spin at room temperature, both become low spin at low temperatures. Some changes of this type may also involve conformational effects – ferric cytochrome *c*, high spin at 100 °C, becomes low spin at room temperature (Keilin, 1966).

Conformational, or chemical, equilibria may contain kinetic steps that are very temperature dependent. In such cases the low temperature spectrum observed will depend upon the rate of freezing. Slow freezing will permit the equilibrium to adjust itself to the value characteristic of the lower temperature; rapid freezing may 'lock in' the situation prevailing at room temperature. Similar results may be obtained on freezing steady state systems; rapid freezing 'traps' the room temperature steady state (Chance and Spencer, 1959); slow freezing modifies the steady state so that, for example, the intramolecular steps in the cytochrome aa_3 reaction become rate limiting, and cytochrome *a* becomes fully reduced and cytochrome a_3 fully oxidized (Jensen *et al.*, 1981; Nicholls and Kimelberg, 1968).

Photolytic reactions, including both photolysis of CO complexes and the initial reactions of photosynthesis, are rather temperature independent. Photolysis of CO complexes of ferrous haem proteins at liquid helium temperatures (Brittain *et al.*, 1982) can be used to prepare low temperature ferrous species whose spectra can be compared with those obtained by freezing samples from room temperature. In this way the existence or otherwise of conformationally different ('out of equilibrium') states can be determined. With appropriate reagents present in the frozen state (obtained by the 'triple trapping' technique) the reactivity of enzymic intermediates can be followed by slowly raising the temperature (Chance *et al.*, 1975).

1.4.5 Medium and Environment

Temperature is just one of many environmental influences upon absorption spectra. Biomolecules that are soluble in more than one type of solvent system will show very different spectral intensities in different solvents. Fig. 13 illustrates this behaviour for chlorophylls and haems. Fig. 13A shows chlorophyll *b* spectra in a variety of organic solvents and in 'solution' in a liposomal suspension (Nicholls *et al.*, 1974). Both the blue and the red absorption bands shift towards the red as the solvent becomes less hydrophobic. A similar transition can be monitored continuously by adding increasing amounts of water to an acetone solution. The liposomal spectrum is both red-shifted and broadened, and the band positions are almost identical with those in chloroplasts. Chlorophyll *a* behaves in a similar way, but its chloroplast spectrum is heterogenous; the liposomal chlorophyll has a 'red' absorption band at 670 nm (Nicholls *et al.*, 1974) which corresponds to that of the shortest wavelength component *in vivo*. Stacking of chlorophyll *a* molecules, and chlorophyll–protein interactions, seem to be involved in producing the species with absorbancy maxima at 685 and 700 nm.

Fig 13B shows haemochromogen spectra obtained at low concentration in largely aqueous phase (Type III) at higher concentrations (Type II) and in 20% pyridine (Type I), as reported by Gallagher and Elliott (1965). Not only is the intensity affected, as with the haematin spectra of Fig. 6, but the peak position in the α-region is also modified by these environmental variations. Protohaemochromogens with nitrogenous ligands in aqueous phase have an α peak at 555–556 nm, moving to 558 nm in excess pyridine and to 563 nm upon aggregation. Such spectral effects may be involved *in vivo* in the differences between different *b*-type cytochromes (Nicholls and Elliott, 1974; Malviya *et al.*, 1980).

In analysing effects of this kind it is important to distinguish between continuous shifts in peak position ('band movement') and equilibria ('band splitting') between two or more distinct chemical species. The latter, but not the former, will be distinguished by the occurrence of distinct isosbestic points (as in Fig. 6).

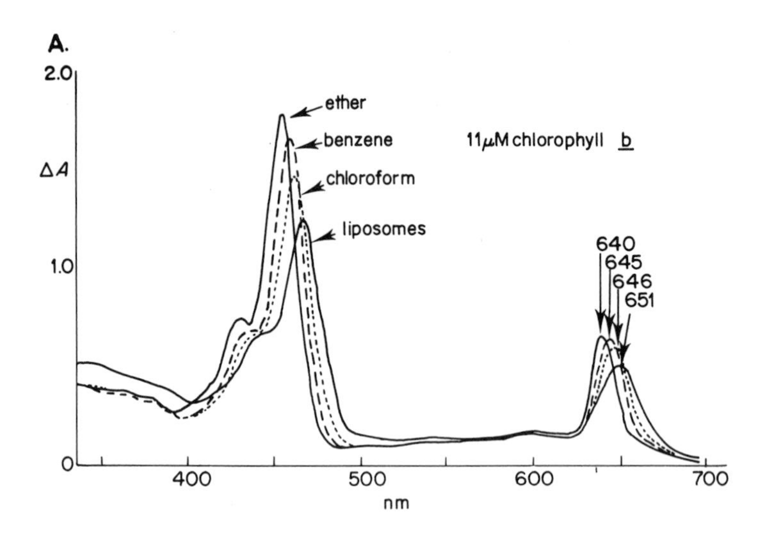

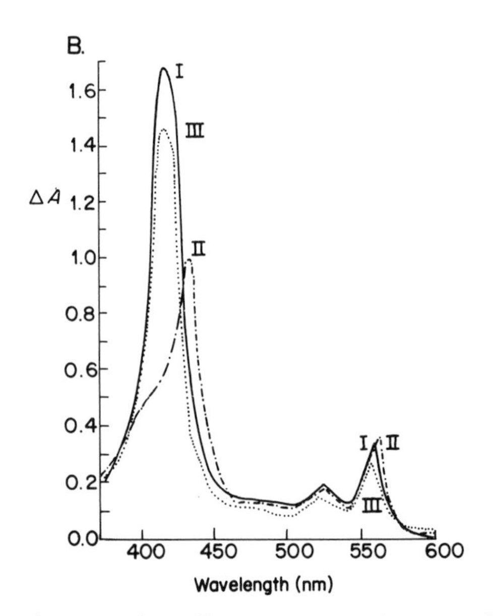

Fig. 13 Solvent and aggregation effects on spectra (see Fig. 6) A. Chlorophyll *b* spectra: 11 μM chlorophyll *b* in diethyl ether (—), benzene (– –), chloroform (- - -) or liposomes (—). Liposomal phospholipids were 96% egg phosphatidylcholine, 4% phosphatidic acid. Solvent as indicated or 20 mM phosphate pH 6.8 (liposomes) 28 °C (reproduced with permission from Nicholls *et al.*, 1974), B. Haemochromogen spectra: 10 μM protohaem (comps. 'I' and 'III') with 1 cm light path or 100 μM protohaem (comp. 'II') with 1 mm light path. Comp. I (—): haematin plus dithionite in 20% pyridine, 20 mM NaOH. Comp. II (–·–): haematin plus dithionite in 0.8% pyridine, 20 mM NaOH. Comp. III (····): haematin plus pyridine (0.1–0.8%) reduced with dithionite in 20 mM NaOH. 25 °C Cary 11 spectrophotometer (reproduced with permission from Gallagher and Elliott, 1965)

In general, aggregation of chromophores has a bathochromic effect, that is, the absorbance per chromophoric group declines (to be distinguished from a 'hypsochromic' effect – a peak movement to shorter wavelengths, as in Fig. 13A). The classical example is the diminished molar absorbance of nucleotides in nucleic acids. Fig. 14 shows that the absorbance of the *E. coli* nucleotides (~ 12,000 M^{-1} cm^{-1} at 260 nm) declines to 9500 M^{-1} cm^{-1} in intact DNA at 95 °C and to less than 7000 M^{-1} cm^{-1} in intact DNA at room temperature. 'Melted' DNA has more absorbance than active DNA, but still less than that of the free nucleotides. The increase on melting DNA ranges from 45% for a pure G:C system to 80% for a pure A:T system (Fig. 14B).

In addition to temperature, solvent and stacking (polymerization) effects, spectra can be modified by placing the chromophore in an electric field. Carotenoids in bacterial chromatophore membranes show spectra that can be shifted when a membrane potential is generated (Jackson and Crofts, 1971). It is uncertain whether this effect represents a true continuous shift in the spectrum or an electrically dependent equilibrium between monomeric carotenoids and polymerized forms.

1.4.6 Band Intensity and Position

Assuming a totally absorbing chromophore, and ignoring all post-Newtonian physics, the maximal molar extinction coefficient (consider the inevitable transparency at high dilutions when only a few molecules are present in the cuvette) may be calculated as (Eq. 3):

$$E_M\,(\max) \simeq \log_{10} E \times N \times 10^{-17}\,\pi r^2 \qquad (3)$$

where N is Avogadro's number and r is the radius of the chromophore molecule in nanometers. When $r = 0.2$ nm, $E_{mM}\,(\max) = 330\ cm^{-1}$. Values of this order of magnitude are approached by the Soret bands of some haem and porphyrin derivatives, examples of which are given in Table 1. Simple double bond systems have absorbance in the ultraviolet region; thus fumarate ($E^{230}_{mM} \sim 4.4$) may be monitored in that part of the spectrum. As the number of conjugate double bonds increases, the absorption maximum moves towards the blue and the intensity increases (see hexaene, item (7) and β-carotene, item (6), with 11 conjugated double bonds in Table 1). The porphyrins (11 conjugated double bonds) and the chlorophylls (10 conjugated double bonds) thus have intense (80–250 mM^{-1}) absorbances in the blue region of the spectrum. The sharper the band, the higher the peak, with the highest values associated with the porphyrins themselves (170–180 mM^{-1}, see item (1) in Table 1) and with the carbon monoxide complexes of the haemoproteins (up to 235 mM^{-1} for the alkaline complex with cytochrome *c*, see item (5) in Table 1).

Aromatic and heterocyclic compounds usually have molar absorption coefficients about an order of magnitude smaller (see riboflavin, item (8) and

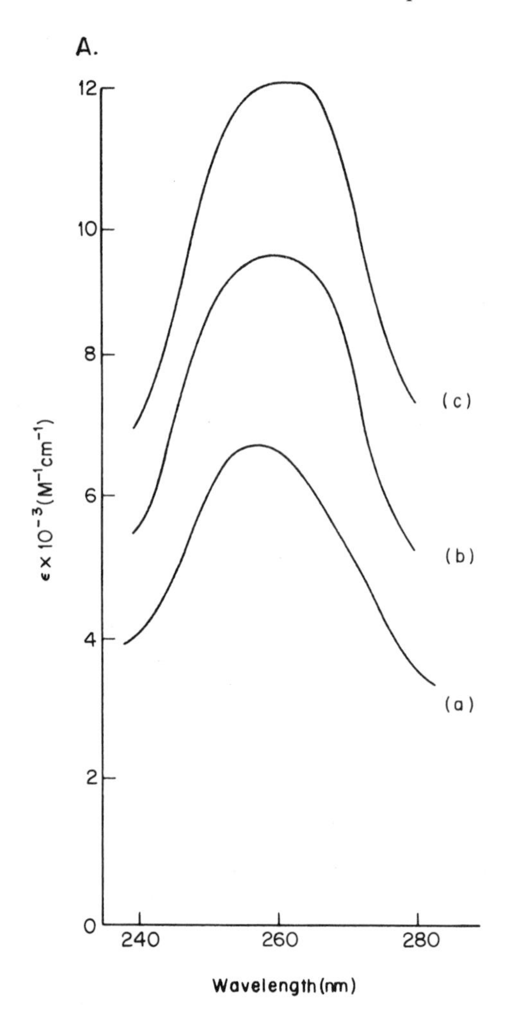

tetracene, item (9) in Table 1). Such species include the nucleotides (Fig. 14 above) and amino acids (Fig. 16 below) with UV bands in the 210–350 nm region. Beyond the Soret (350–450 nm) region we come to the visible region of the spectrum (400–700 nm), dominated by absorption bands of (1) heterocyclic dyes and organic free radicals and (2) transition metal complexes. The absorbances involved are characteristically 1/5 to 1/10 those seen in the Soret region. Thus protoporphyrin IX has a series of four sharp bands between 500 and 630 nm ranging from 15 $mM^{-1}cm^{-1}$ to 7 $mM^{-1}cm^{-1}$ in intensity; the two bands of oxyhaemoglobin at 542 and 577 nm have E_{mM} values of 14.4 and 15.3 $mM^{-1}cm^{-1}$ respectively. The most intense bands in this region are of the haemochromogen or cytochrome type, e.g. ferrocytochrome c ($E = 28.0$ mM^{-1}

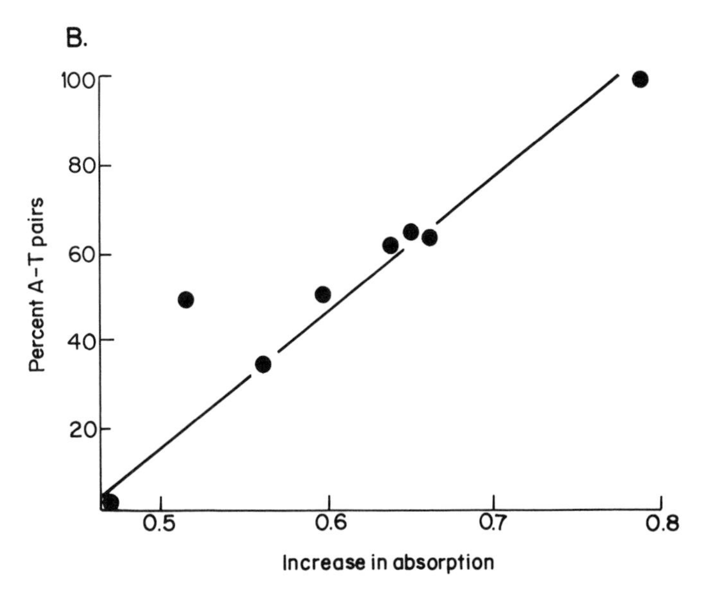

Fig. 14 Effect of intramolecular interactions upon spectra: DNA hyperchromism. A. *E. coli* DNA in 50% methanol: (a) room temperature (native) spectrum; (b) 95 °C spectrum (denatured); (c) spectrum of component deoxyribonucleotides (complete hydrolysis). B. Correlation of hyperchromic effect with proportion of adenine–thymine pairs in different DNA samples. Increase in absorption on treatment at 95 °C is plotted against percentage of A–T pairs. (Brown, 1980; after Mahler *et al.*, 1964). Reproduced with permission from Brown (1980). *An Introduction to Spectroscopy for Biochemists.* Copyright: Academic Press Inc. (London) Ltd

cm^{-1} at 550 nm) or pyridine protohaemochromogen, $E = 33.0$ $mM^{-1}cm^{-1}$ at 558 nm (Fig. 13B).

For instrumental reasons (the absence of lead sulphide detectors on many instruments) the near-infra red region (700–2000 nm) has been less studied. Absorption bands in this region are commonly even lower than in the visible region. The bacterial chlorophylls are the most important group of biological pigments with absorbance in this part of the spectrum. With an extra saturated bond in the resonating ring structure, the α peak moves out into the 750–850 nm region. This is presumably an adaptive trait to permit utilization of the far red light that escapes Rayleigh scattering to penetrate the depths of the lakes and seas in which the purple and green sulphur bacteria live. Haemoglobin complexes have characteristic, if weak, absorbancies in this region (Lemberg and Legge, 1949) and the recent development of near IR magnetic circular dichroism (MCD) measurements (Eglinton *et al.*, 1980) has shown that these bands may be of diagnostic importance. Measurements are complicated by the absorbancies of water (1500–2000 nm) and deuterium oxide ($>$ 2000 nm).

Certain copper complexes, including at least one of the two copper centres in

Table 1 Soret bands of porphyrins and haem compounds, compared with maximal absorbances of other organic conjugated species

Compound	Soret maximum	E ($mM^{-1}cm^{-1}$)	Ref.
(1) Protoporphyrin IX dimethyl ester (in $CHCl_3$)	407 nm	171	(a)
(2) Oxyhaemoglobin	415 nm	125*	(b)
(3) Carbonmonoxy haemoglobin	419 nm	194*	(b)
(4) Chlorophyll *a* (in ether)	420 nm	111	(c)
	660 nm	84	(c)
(5) Carbonmonoxy ferrocytochrome *c* (pH 14)	415 nm	235	(d)
(6) β-Carotene (all-*trans*)	450 nm	150	(e)
(7) $H(CH{=}CH)_6H$ (in iso-octane)	364 nm	138	(a)
(8) Riboflavin	445 nm	12	(a)
(9) Tetracene (in benzene)	445 nm	10	(a)

* per subunit.
Refs. (a) Brown (1980)
(b) Waterman (1978)
(c) Sauer (1975)
(d) Keilin (1966)
(e) Jaffe and Orchin (1962)

cytochrome *c* oxidase, also have characteristic low absorbancies in this region (Malkin and Malmström, 1970). The so-called 'blue' copper proteins all have a pair of characteristic absorption bands, the first in the red part of the spectrum (597 to 625 nm) with $E = 3.5$–5.0 $mM^{-1}cm^{-1}$, the second in the near IR region (760–845 nm) with $E = 0.4$–1.8 $mM^{-1}cm^{-1}$, per copper atom. The former absorption band may be confused with those of haem proteins (especially cytochromes of the *a* type) although it is characteristically much broader than a haem α band. The 'non-blue' copper proteins lack these bands. One of the copper atoms in cytochrome *c* oxidase shows an 830 nm band ($E \sim 1$–2 $mM^{-1}cm^{-1}$), but lacks a well-defined band in the 600 nm region, and therefore cannot be simply classified as 'blue' or 'non-blue'. It may be noted that these absorption bands are present in the cupric state, and abolished upon reduction, in contrast to the characteristic haem absorptions, which usually become more marked on reduction (see section 1.5.5).

1.4.7 Kinetic Measurements

Following optical changes at a single wavelength (or wavelength pair) as a function of time is probably the commonest spectrophotometric technique. A small amount of enzyme or other reaction-initiating reagent is added to a larger volume of substrate in a cuvette and the subsequent increase or decrease in absorbance is monitored continuously. Fig. 15A illustrates the principle of the method. Some problems that may be encountered include the following:

(1) dilution effects,
(2) activation and deactivation phenomena,
(3) uncertainty of reaction order,
(4) uncertainty of end-point.

These problems are interrelated. If dilution of the material in the cuvette is minimal, dilution of the reagent added will be considerable. This may modify its activity in a time-dependent way, which in turn will alter the apparent kinetics of the reaction time course.

To minimize perturbation of the initial mixture in the cuvette (3 ml) the initiating enzyme or reagent should be added in a volume of 0.1 ml (3% total) or less. If a *decrease* in A is being followed, this initial dilution effect should be taken into account in calculating A_0. Volume changes, even at the 3% level, may not be trivial. Yeast isocitrate dehydrogenase, a complex allosteric enzyme, can be placed in solution under conditions in which its responses to isocitrate, NAD^+, Mg^{2+} and ATP are all power functions, so that if none is at a saturating level, the reaction rate is proportional to the eleventh power of the total concentration. A 3% dilution would then include a $1 - (0.97)^{11}$ decline in rate, or nearly 30%.

The reagent being added will experience a thirty-fold or greater dilution into a medium at least in one respect different from that in which it was dissolved. When the reagent is the enzyme this can result in activation or deactivation events which will be seen as time-dependent increases or decreases in reaction rate. If a reaction is zero order, a deactivation effect superimposed may make it look first order; vice versa, if a reaction is first order, an activation effect may make it approach zero order characteristics. The test must be to initiate the reaction in some other way, e.g. by substrate addition to prediluted enzyme. But if the process of (de)activation involves contact between enzyme and substrate, even this will not discriminate. A second pulse of substrate after the first addition has been 80–90% converted may check this point.

If the initial reaction order still cannot satisfactorily be established, then more complex plotting procedures than A *vs* t or log A *vs* t must be used. For example, progress curves can be plotted in the form $t/\ln(S_0/s)$ *vs* $(S_0 - s)/\ln(S_0/s)$ where t is time, S_0 = initial substrate concentration and s = substrate concentration at time t. Such curves have a slope of zero if the kinetics are first order. If the reaction is following Michaelis–Menten kinetics and there is no product inhibition, then the plot will approximate a Hanes plot (S_0/v *vs* S_0) for the reaction under consideration (Cornish-Bowden, 1980; Brooks and Nicholls, 1982).

If a reaction is known to be first order, yet its end-point is uncertain, one of the plots recommended by Guggenheim (1926) or by Kezdy *et al.* (1958) may be used (see Cornish-Bowden, 1980). The Guggenheim method plots log $(A_t - A_{t+\Delta t})$, the logarithm of the difference between the absorbance at time t and

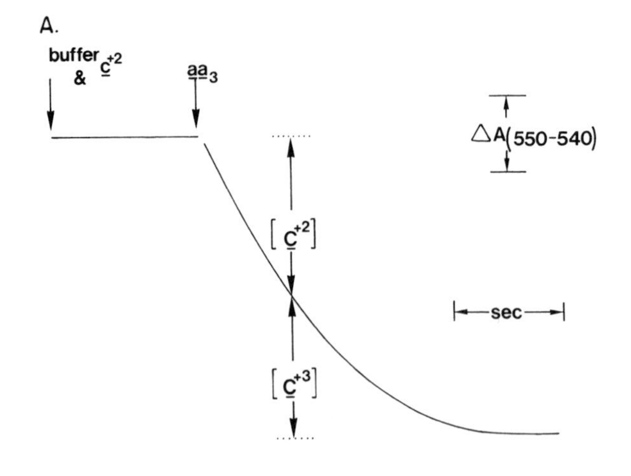

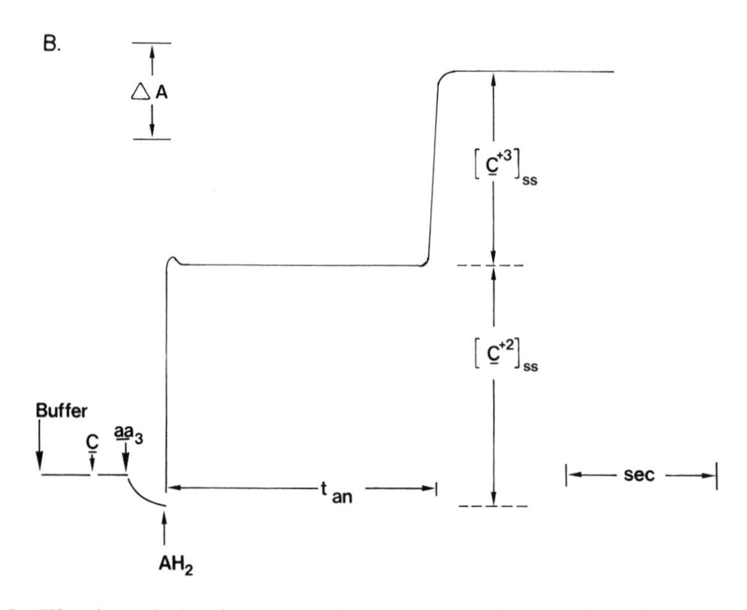

Fig. 15 Kinetic analysis using spectrophotometry. A. Slow transient kinetics: oxidation of cytochrome *c* by cytochrome aa_3. A small volume of reagent (enzyme, aa_3) added to a solution of substrate (c^{2+}, ferrocytochrome *c*) and the subsequent oxidation ($c^{2+} \rightarrow c^{3+}$) monitored at 550 nm (or 550–540 nm) as a function of time ($t_{1/2} \geqslant 5$ seconds). B. Slow steady state kinetics: oxidation of ascorbate plus cytochrome *c* by cytochrome aa_3. After addition of aa_3 to aerobic buffer containing *c*, a steady state reduction is obtained by adding ascorbate (AH_2), which is maintained ($[c^{2+}]$ss) until the exhaustion of oxygen from the system at the anaerobiosis point (t_{an}). C. Fast transient kinetics: stopped flow. Anaerobic mitochondria are mixed with oxygen in a regenerative stopped-flow instrument and the changes at 605–615 nm (cytochrome *a*, upper trace) monitored. The oxidation during flow and the subsequent full oxidation on stopping the reaction are shown (the rate constant shown was evaluated from the short period of 'continuous

C.

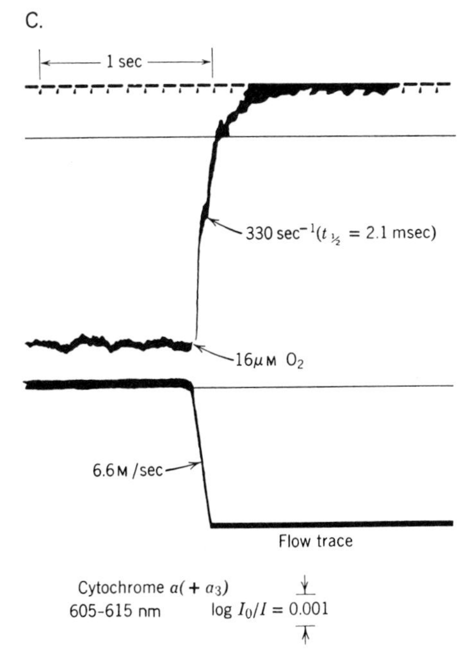

D.

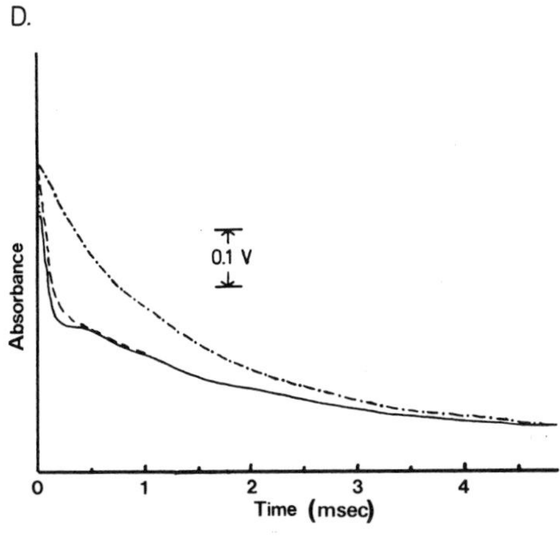

flow'; a somewhat slower rate would be obtained in this experiment from the time course after the 'stop'). The lower trace indicates piston position and therefore monitors flow rate (reproduced with permission from Chance *et al.*, 1965). D. Fast transient kinetics: flash photolysis. Absorbance changes at 446 nm during oxidation of fully reduced cytochrome aa_3 (in solution) at three levels of oxygen (-·-·-, 17 μM; – –, 274 μM; —, 720 μM) at pH 7.4 and 13 °C using a flash photolysis instrument similar to that in Fig. 4 (time constant ~0.05 msec) but attached to flow system as in Fig. 3. After mixing with O_2-containing buffer the CO-inhibited reduced enzyme (2 μM aa_3 + 0.5 mM CO) is activated by photolytic removal of CO. Following photolysis, oxidation occurs with a half-time of 1 msec (-·-·-) or faster; only with an instrument responding in less than 1 msec can the intermediate state(s) obtained with higher O_2 levels, (– –, —) be observed (Hill, 1982; courtesy Dr. B. C. Hill) that a few seconds later (Δt), against t. The result is a straight line with slope equal to k, the first order rate constant. The Kezdy *et al*. method plots A_t *vs* $A_{t+\Delta t}$ for a series of values of t and a constant value of Δt; the slope of this plot is $e^{k\Delta t}$ and the A_∞ value is that obtained when the plot intersects with the plot for $\Delta t = 0$.

Steady state analysis (Fig. 15B) can only be used with certain types of enzyme system. When available, it provides a very useful adjunct to non-steady state (depletion) methods such as that in Fig. 15A. If an enzyme or associated cofactor (cytochrome c in Fig. 15B) can exist in a spectroscopically identifiable intermediate state during a reaction, both the overall reaction rate and some intermediate steps can be analysed. Fig. 15B represents the pattern of absorbance change at 550 nm (the α-maximum of ferrocytochrome c) in a single beam spectrophotometer or at 550–540 nm in a double beam (dual wavelength) instrument. Cytochrome c is reduced by a substrate such as ascorbate, according to Eq. 4:

$$\mathrm{AH} + c^{3+} \rightarrow \mathrm{A} + c^{2+} + \mathrm{H}^{+} \tag{4a}$$

$$v = k_r\,[c^{3+}]\,[\mathrm{AH}] \tag{4b}$$

and reoxidized by the enzyme cytochrome c oxidase according to Eq. 5:

$$c^{2+} + e \rightleftharpoons c^{2+}e \xrightarrow{O_2\&c} c^{3+}\,\mathrm{e} \rightleftharpoons \mathrm{e} + c^{3+} \tag{5a}$$

$$v = k_0\,[\mathrm{c}^{2+}]e/(K_m + [c]_{\mathrm{total}}) \tag{5b}$$

the kinetic form of which gives rise to Smith-Conrad (first order) kinetics as analysed by Minnaert (1961).

In the steady state condition, where the redox state of c does not change, the velocities of 4b and 5b are equal, and the steady state redox ratio is given by Eq. 6:

$$\text{steady state } \frac{[c^{2+}]}{[c^{3+}]}_{(\mathrm{ss})} = \frac{k_r\,[\mathrm{AH}]\,(K_\mathrm{m} + [c]_{\mathrm{total}})}{k_0 e} \tag{6}$$

When total $[c]$ is varied, the steady state will vary and a plot of $v/[\mathrm{AH}]$ against $[c^{3+}]_{\mathrm{ss}}$ will give a value for k_r, the rate constant for reduction. At the same time, a plot of $[c^{2+}]_{\mathrm{ss}}/v$ against $[c]_{\mathrm{total}}$ will give a line with slope $1/k_0e$ and (negative) intercept equal to $-K_\mathrm{m}$ on the $[c]_{\mathrm{total}}$ axis. The velocity, v, is constant during the steady state (see the O_2 uptake trace in Fig. 11, Chapter 2) because the affinity of the enzyme for oxygen is very high. Provided that the oxygen concentration is known, the velocity in electron equivalents per litre is given by Eq. 7:

$$v = 4\,[O_2]/t_{\mathrm{anaerobic}} \tag{7}$$

A similar type of analysis can be made with steady state cycles in which the substrate, rather than acceptor, is the limiting species. NADH, with a high affinity for the NADH dehydrogenase, can be used with submitochondrial particles and cytochrome *c* to cycle the latter through the partially reduced steady state back to the fully oxidized form (rather than the final fully reduced form seen in Fig. 15B). Similar kinetic expressions can be used to those derived by Chance (1943) for analysing the cycles of peroxidase–peroxide complexes.

Steady state measurements can be affected by the same problems of activation and deactivation of the enzyme seen with the direct depletion method. In addition, the initial phases of the steady state may be affected by the 'mixing relaxation' phenomenon. Consider the concentration of reductant seen by the initially oxidized cytochrome *c* upon mixing (AH_2 arrow). As the reductant is diluted out after addition, the nearest molecules of *c* experience a level much greater than the final steady state concentration; all the cytochrome *c* experiences an initial level in excess of the final level. If the reduction rate is faster than the mixing rate, the percentage reduction will start out higher than the final steady state to which it will then relax; compare the initial 'blip' preceding the steady state in Fig. 15B.

A key feature of the stopped flow instrument (Fig. 3) is thus the mixing chamber. In many cases once the initial mixing has been achieved the resulting reaction rate can be monitored using conventional recording methods. The first observations on compound I of catalase, the catalytic intermediate in the decomposition of H_2O_2, required rapid mixing in order to avoid the decomposition of the added H_2O_2 by that small part of the enzyme with which it initially came into contact. The subsequent steady state is long-lived; indeed, the intermediate only decays in the presence of hydrogen donors other than H_2O_2 (Chance, 1947).

Because of the difficulty of chopping the beam at a sufficient rate ($>10^3$ cycles/sec), most stopped flow instruments are of the single beam variety, although the introduction of diode array machines in this area may be changing the situation. Fig. 15C illustrates typical traces obtained from a stopped flow device such as that indicated schematically in Fig. 3. Mixing occurs within 2 msec in this apparatus, eliminating the 'mixing relaxation' problem for all except very fast bimolecular processes. But flow times of up to 15 msec limit the data acquisition for some of the faster reactions. Conventional storage oscilloscopes triggered by the 'stop' signal may lose the optical information available prior to stopping the reaction (and seen in Fig. 15C). Continuous memory devices avoid this problem but the solution stability during the flow period may be inadequate to obtain rate information during that time (see discussion in Hiromi, 1979). First order rate constants of about 250 sec^{-1} are about the maximum that can be reliably examined; because extinction coefficient changes rarely exceed 50 $mM^{-1}cm^{-1}$, this limits measurable second order rate constants (at 1 μM concentrations) to $\leq 2 \times 10^8\ M^{-1}sec^{-1}$, some 10 to 100-fold smaller than some diffusion-limited rates.

To examine rates with half times less than 2 msec ($k \geq 500\ sec^{-1}$), it is necessary to abandon mechanical mixing for other ways of starting reactions.

Fig. 15D illustrates the time courses for the reactions of reduced cytochrome *c* oxidase with molecular oxygen. The reduced enzyme is prepared as the CO complex, mixed with oxygenated buffer in a flow device, and the CO is then photolysed with a suitable flash apparatus. The reaction complexities are only seen at high oxygen levels, for which the flow apparatus alone would be quite inadequate. But the method is limited to the study of those reactions one of whose starting species can be generated photochemically. The commonest such reaction is the preparation of ferrous haem proteins by the photolysis of their CO complexes. However photochemical reductions are also known (Nicholls and Chanady, 1981) and the use of intense laser flashes (Ahmad *et al.*, 1982) may make it possible to study reductive as well as oxidative processes by this technique.

Both practical and theoretical (kinetic) aspects of stopped flow and other rapid spectrophotometric methods are discussed by Hiromi (1979).

1.5 Biological Examples

1.5.1 Respiratory Organelles

Since the original spectroscopic studies of Keilin (1925), optical methods have predominated in the study of these organelles and their enzymes. Eukaryotic mitochondria, and the submitochondrial particles derived therefrom, contain three major types of component with visible spectra, the cytochromes, the flavoproteins, and the iron–sulphur proteins, as well as a quinone system (coenzyme Q) with UV absorbance.

A small catalogue of characteristic spectra was published in the review article of Nicholls and Elliott (1974). Table 2 lists some characteristic wavelengths and extinction coefficients for the components involved. The microsomal fraction also contains cytochromes b_5 and P450, and the peroxisomal fraction contains catalase and flavoprotein oxidases although no spectrophotometric studies have so far been carried out with intact peroxisomes.

To eliminate the problems of light scattering (Fig. 10A), difference spectra of reduced minus oxidized samples (Fig. 10B) are obtained. With submitochondrial particles, total reduction is usually achieved by the addition of a small amount of sodium dithionite, which both reduces the redox components in the membrane and depletes the system of oxygen. A few minutes should be allowed between dithionite additions and measurement. The fully oxidized sample (placed in the reference cuvette) can be obtained by simply allowing the submitochondrial particles to remain aerobic in the absence of added substrates. An alternative reduced sample can be prepared by adding a natural substrate in excess (usually succinate or NADH) and allowing the system to become anaerobic. The difference between this and the dithionite-reduced sample represents the redox components not accessible to substrate.

Table 2 Extinction coefficient and wavelengths of maximal absorbance for the major cytochromes, flavoproteins and related pigments

Component	Spectrum	λ_{max} (nm)	E (mM^{-1}cm^{-1})	Ref.
Cytochrome aa_3‡	Red–ox	605	24.0*	(a)
Cytochrome b	Red–ox	562	~18.0	(a)
Cytochrome c_1	Red–ox	553	~20.0	(a)
Cytochrome c	Red–ox	550	20.4	(a)
Cytochrome P450	Red CO–red	450	85†	(b)
Lipoate dehydrogenase	Ox	455	11.0	(c)
L-Amino acid oxidase	Ox	460	11.0	(c)

* 27.0 mM^{-1}cm^{-1} for $\Delta\Delta E_{mM}$ (605–630 nm, reduced–oxidized)
† 91 mM^{-1}cm^{-1} for $\Delta\Delta E_{mM}$ (450–490 nm, red CO–red)
‡ haems/functional unit
Refs. (a) Nicholls and Elliott (1974)
(b) Omura and Sato (1964); Sato and Omura (1978)
(c) Penzer (1980)

If an 'absolute' spectrum is required, some methods must be found for correcting the light scattering of the sample. Diluted milk, which shows both Rayleigh and Tyndall scattering components (Fig. 8), can often be used. Alternatively, the pigments in the particles can be destroyed by incubation with a bleaching reagent such as peroxide (examples are given in Nicholls and Elliott, 1974) and the resulting difference spectrum will represent the absolute spectrum of the pigment mixture itself.

Keilin and Hartree (1958) pioneered two methods for dealing with such systems:

(1) correcting for scattering by placing the sample at varying distances from the photomultiplier tube and extrapolating to zero distance;
(2) suspending the sample in a medium whose refractive index is close to that of the particles themselves.

A third method is to disperse the membranes using an appropriate detergent system. For chloroplast and plasma membranes, 1% Triton X-100 may be used, but for mitochondrial membranes 1% sodium cholate is preferred. Cytochromes aa_3 and c_1 (as well as the already soluble cytochrome c) are apparently spectroscopically unaffected by this treatment, although the catalytic activity of cytochrome aa_3 (cytochrome c oxidase) is inhibited.

Cytochrome b of mitochondria is composed of two components, readily distinguished spectroscopically in intact mitochondria and in submitochondrial particles with intact membranes (Nicholls and Elliott, 1974; Malviya *et al.*, 1980). On treatment with detergents, however, at least one of the two components is modified and the two become spectroscopically indistinguishable. Cytochrome 'b_K' is characterized *in situ* by an α-maximum of a symmetrical kind centred at 562 nm, and an energy-state independent redox potential of +60 mV at pH 7 and 25 °C; cytochrome 'b_T' *in situ* has a split α-maximum with the main peak at 565 nm and a shoulder at 558 nm, and an energy-state dependent redox potential of about −60 mV in de-energized systems and +240 mV in energized systems. Isolated cytochrome *b* has a single peak at 563 nm and a redox potential more negative than either value measured *in situ*; it does, however, possess two haem groups associated with a single polypeptide chain; Erecinska *et al.* (1973) have monitored the progressive spectroscopic transformation of one or both haems during detergent (chaotropic) treatment.

The most striking example of this 'allotopic' behaviour (Racker, 1976) is that of the hydroxylase system, P450, of microsomal (and some mitochondrial) membranes. On detergent isolation the unusual CO-induced spectrum with γ-peak at 450 nm and 'high-spin' visible spectrum is transformed into a conventional CO mixed haemochromogen spectrum with α-peak at 420 nm and typical 'low-spin' visible spectrum – the P450 to P420 conversion. This made it essential to use physical techniques to correct for light scattering in reporting absolute spectra, although more recently true soluble P450 preparations have been obtained.

Difference spectra (reduced *minus* oxidized) can be used to obtain immediately the concentration of the several cytochromes, for their spectroscopic maxima are clearly separated. To do the same for the iron–sulphur flavoproteins present is more difficult, because their UV and visible absorption bands overlap. However, some information can be obtained by the use of specific inhibitors. Thus malonate addition to submitochondrial particles produces a characteristic bleaching presumably due to succinate dehydrogenase (Nicholls and Malviya, 1968); and reduction by NADH in the presence of rotenone presumably shows only the NADH dehydrogenase. But more detailed study requires the use of other techniques, especially fluorescence and EPR analysis.

1.5.2 Proteins and Amino Acids

All proteins have UV end absorption associated with the peptide bond; the circular dichroism in this region has been used to analyse protein structure. Absorbance spectroscopy, however, is usually confined to the 280 nm region, in which only the three aromatic amino acids show absorption bands, as illustrated in Fig. 16A. At neutral pH, tryptophane has the highest molar absorbance with a three-banded spectrum with maximum at 280 nm and shoulders at about 272 and 287 nm; tyrosine shows a split peak at 274 and 280 nm which shifts to 290 nm and increases in intensity upon ionization, while phenylalanine has a very low absorbance but with a characteristic five-banded structure centred at 257 nm (Table 3). Can these quite different spectral behaviours be used to determine the aromatic amino acid content of different proteins? Not quite. However, the relative amino acid composition can be determined by

Table 3 Amino acid absorption peaks and extinction coefficients

Amino acid	Chromophore (R-group)	λ_{max} (nm)	E (mM^{-1}cm^{-1})
Cysteine	$-S^-$	235	3.20
Cystine	–S–S–	250	0.32
Phenylalanine	Phenyl	257	0.22
Tyrosine	Phenol	274	1.44
Tryptophane	Indole	280	5.05

(Values for pH 6–7, room temperature)
Ref. Brown (1980), see Fig. 16A

inspection fairly easily. Table 4 compares the amino acid compositions of four proteins, one high in phenylalanine (serum albumin), one in tyrosine (trypsin), one in tryptophane (chymotrypsinogen) and one low in all aromatic residues (gelatin/collagen).

The corresponding spectra are seen in Fig. 16B. The broad absorption typical of tyrosine (trypsin) is clearly different from the sharp bands characteristic of the other species, with the long wavelength shoulder diagnostic of tryptophane (chymotrypsinogen) and the shorter wavelength five-banded pattern of phenylalanine (serum albumin) also seen in the first derivative spectrum (Fig. 11). This latter spectral type is exceptionally well developed in the protein calmodulin (Jarrett and Penniston, 1978) which has up to ten phenylalanine residues associated with only two tyrosines and no tryptophane. If, however, a more quantitative analysis is attempted, it is found that some of the residues must have absorption peaks shifted or modified compared with those of the free amino acids. As with the nucleic acids (Fig. 14) 'stacking' and hydrogen bonding interactions may occur in the interior of the protein molecules where many of these residues are located. The chymotrypsinogen spectrum seems to be somewhat red-shifted compared to that of free tryptophane; and the intensity of the trypsin spectrum is greater than that which would be predicted from the sum of the absorbances of the amino acid

Table 4 Aromatic amino acid contents of four proteins (see Fig. 16B)

Protein	% phenylalanine	% tyrosine	% tryptophane
(a) Bovine serum albumin*	6.4	3.5	0.2
(b) Porcine trypsin*	1.8	3.6	0.9
(c) Chymotrypsinogen*	2.3	1.5	3.0
(d) Gelatin*	<1.0	<0.5	<0.1
(e) Calmodulin†	4.8–7.0	1.3	0.0

* Spectrum in Fig. 16B
† Spectrum in Fig. 3 of Jarrett and Penniston (1978)

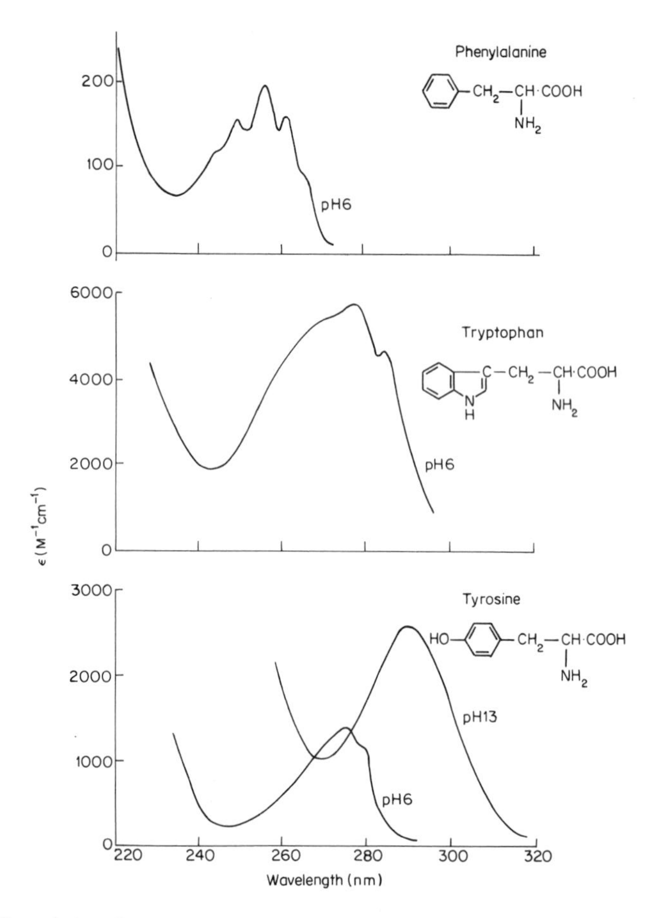

Fig. 16 Ultraviolet absorption spectra of amino acids and proteins. A. UV spectra of aromatic amino acids. Phenylalanine and tryptophane at pH 6, tyrosine at pH 6 and pH 13 (tyrosinate): see Table 3 (Brown, 1980). Reproduced by permission from Brown (1980), *An Introduction to Spectroscopy for Biochemists*. Copyright: Academic Press Inc. (London) Ltd. B. UV spectra of four proteins: 1.33 mg ml^{-1} bovine serum albumin (see Fig. 11); 0.42 mg ml^{-1} chymotrypsinogen; 0.70 mg ml^{-1} trypsin; 1.00 mg ml^{-1} gelatin; see Table 4 (Nicholls, 1982)

residues it contained. However, it is clear that qualitative use of these spectral differences in differentiating between proteins and monitoring purity is quite feasible.

More detailed structural information can be obtained by 'perturbation' methods (Donovan, 1973; Brown, 1980). Solvent perturbation spectra are obtained by taking the difference spectra in the presence and absence of, say, high (20%) levels of glycerol, sucrose or ethylene glycol. Similarly acid denaturation induces changes of the tyrosine and tryptophane spectra as they are

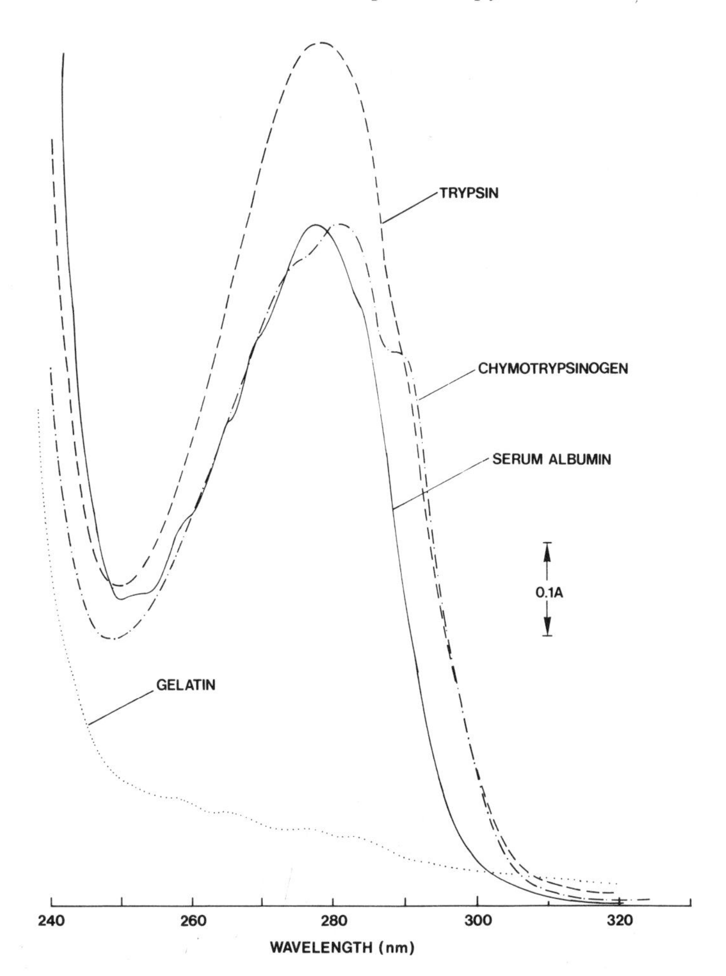

exposed to the medium rather than the protein interior; exposure produces a bathochromic effect. Denatured peptides have absorption spectra red-shifted compared with those of the free amino acids (Brown, 1980). These effects therefore contrast with those seen in nucleic acids (Fig. 14) – denaturation producing a bathochromic rather than a hyperchromic effect, and hydrolysis producing a blue shift rather than a further hyperchromic effect.

1.5.3 Nucleic Acids and Nucleotides

For a given type of nucleic acid, the absorption is approximately constant, as all the residues have approximately the same extinctions. As the purine nucleotides have somewhat higher absorbances than the pyrimidine nucleotides, and as stacking in duplex DNA is presumably more constant than internucleotide relationships in RNA, the former should have the more constant of the two absorbances (Table 5). Absorbances are, however, very pH dependent (Beavan *et al.*, 1955), both for the free nucleotides and for intact DNA or RNA. In the case of the intact nucleic acid, the pH variation may be increased by the hyperchromic effect (Fig. 14) if high or low pH transitions modify the H-bonding and hence the internucleotide interactions.

Table 5 Extinction coefficients at wavelengths of maximal absorbance for nucleotides and nucleic acid

Nucleotide or nucleoside	pH	λ_{max} (nm)	E ($mM^{-1}cm^{-1}$)
Adenosine	6.4	260	14.9
Guanosine	6.0	253	13.7
Uridine	7.2	262	10.1
Thymidine	7.2	267	9.65
Cytidine	7.2	271	9.1
Calf thymus DNA	7.0	257	6.8*
Calf thymus DNA	2.0	262	8.7*
Calf thymus DNA	12.3	260	8.2*

Ref. Beavan *et al.* (1955)
* per residue

1.5.4 Coenzymes

Although the traditional extinction coefficient of 6.22 $mM^{-1}cm^{-1}$ at 340 nm for reduction of NAD^+ to NADH is still the one commonly used, a careful and more recent study recommends the rounded value of 6.3 $mM^{-1}cm^{-1}$ at 339 nm for both NADH and NADPH (Table 6), at a standard temperature of 30 °C. Temperature sensitivity (Malcolm, 1973) is one of the commonly neglected features that leads to the use of incorrect molar absorbances. It should also be noted that if the oxidized form (with adenine nucleotide absorbing in the 260 nm region) has any end absorbance at 340 nm, then the absolute and difference spectrum values will not be identical.

The reduced forms (NADH and NADPH) are 'cation' (including H^+) sensitive, while the oxidized (NAD^+ and $NADP^+$) are 'anion' (including OH^-) sensitive; the differential acid and alkali sensitivity forming the basis of some of Lowry and Passonneau's microanalytical techniques (1972). The combination of anions with NAD^+ (or $NADP^+$) gives rise to complexes with increased absorbance at 340 nm, which can be mistaken for reduction. Cyanide, a commonly used inhibitor of NADH reoxidation, readily forms such a complex with NAD^+; in the presence of cyanide, therefore, apparent reduction

Table 6 Extinction coefficients and wavelengths of maximal absorbance for NADH and NAD^+

λ (nm)	E ($mM^{-1}cm^{-1}$) values	
	NAD^+	NADH
260	18.0	15.0
330†	<0.1	~6.0
334	<0.1	6.11 (NADPH, E_{mM} = 6.13)
339 (340 ?)*	<0.05	6.3 (IFCC recommendation)
365	<0.05	3.39 (NADPH, E_{mM} = 3.45)

† isosbestic for temperature variation from 15 to 35 °C
* temperature-dependent: $\Delta E_{mM}/\Delta T = -8 \times 10^{-3}$ A/°C
Refs. Malcolm (1973), Netheler (1974)

monitored at 340 nm should be checked by scanning the spectrum of the product, which will distinguish the reduced from the cyan forms.

1.5.5 Haemoglobins

Haemoglobins and related compounds can be identified spectrophotometrically in one of several ways. One of the oldest, yet still the most sensitive method is to form a haemochromogen species (see Fig. 13B) by denaturing the protein at pH 13 (0.1 M NaOH), adding 10–20% pyridine or other nitrogenous ligand, and reducing the mixture with sodium dithionite (the ferric form, or 'parhaematin' has a less sharp and characteristic spectrum). The resulting species (compound II in Fig. 13B), with α-band at 558 nm, is almost immediately diagnostic of a protohaem compound. The cytochromes were discovered as naturally occurring haemochromogens, but their haem groups were initially classified into *a*, *b*, and *c* types by the position of the α-bands of the corresponding pyridine haemochromogens.

Cytochromes of the '*a*' type form haemochromogens with α-peaks in the 580–600 nm region; cytochromes of the '*b*' type form haemochromogens with α-peak at 558 nm (protohaemochromogen); cytochromes of the '*c*' type form haemochromogens with α-peak at 550 nm (haematohaemochromogen-type). Since the original classification was developed, the '*d*' cytochromes have been distinguished from the '*a*' type by their chlorin haems with haemochromogen bands beyond 600 nm. And a class of partial *c*-type haems has been identified with only one of the two vinyl groups attached to a protein cysteine, and these show a haemochromogen α-peak at 553–554 nm, midway between the *c* and *b* (protohaem) forms.

The spectra of native ferrous haemoglobin complexes are shown in Fig. 17A. Unliganded haemoglobin, and all similar haem proteins, contains a pentacoordinate iron atom and shows a single broad band in the visible region. When

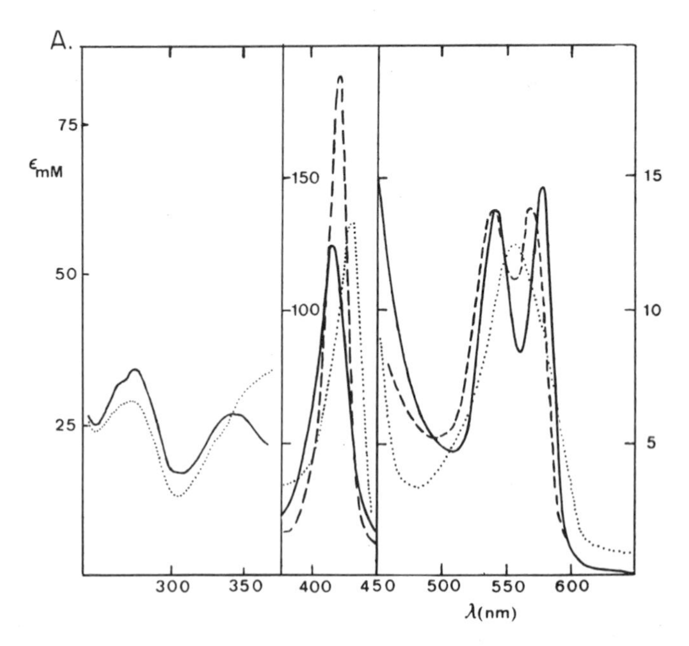

Fig. 17 Haemoglobin spectra: standard and polarized. A. Absorption spectra of solutions of human haemoglobin in UV, Soret and visible regions, at 20 °C in 0.1 M phosphate, pH 7.0: ---- deoxyhaemoglobin; — oxyhaemoglobin; – – carbonmonoxy-haemoglobin (Waterman, 1978). B. Polarized absorption spectra of oxyhaemoglobin crystals in Soret, visible and near IR regions. Measuring light plane polarized parallel to crystal b axis (upper spectrum) or *a* axis (lower spectrum). Upper plot (x/y) indicates ratio of absorbances as a function of wavelength (Eaton and Hofrichter, 1981). C. Polarized absorption spectra of deoxyhaemoglobin crystals in Soret, visible and near-IR regions. Measuring light plane polarized parallel to crystal C* axis (upper spectrum) or *a* axis (lower spectrum). Upper plot (x/y) indicates ratio of absorbances as a function of wavelength (Eaton and Hofrichter, 1981). Reproduced with permission

ligated by a gas molecule (CO or O_2) the visible region shows a split peak with almost equal α- and β-bands (unlike the haemochromogens with α-peaks much higher than β-peaks). The Soret (γ) band of the CO complexes is much higher than the others (see Table 1 and discussion). The λ_{max} positions and extinction coefficients are listed in Table 7.

Haemoglobins in the ferric state (methaemoglobins) are functionally inert. Their spectra are quite distinct from those of the ferrous forms and fall generally into two classes:

(1) the high-spin type with charge transfer bands above 600 nm and in the 480–500 nm region (e.g. acid (neutral) methaemoglobin and its fluoride complex); and

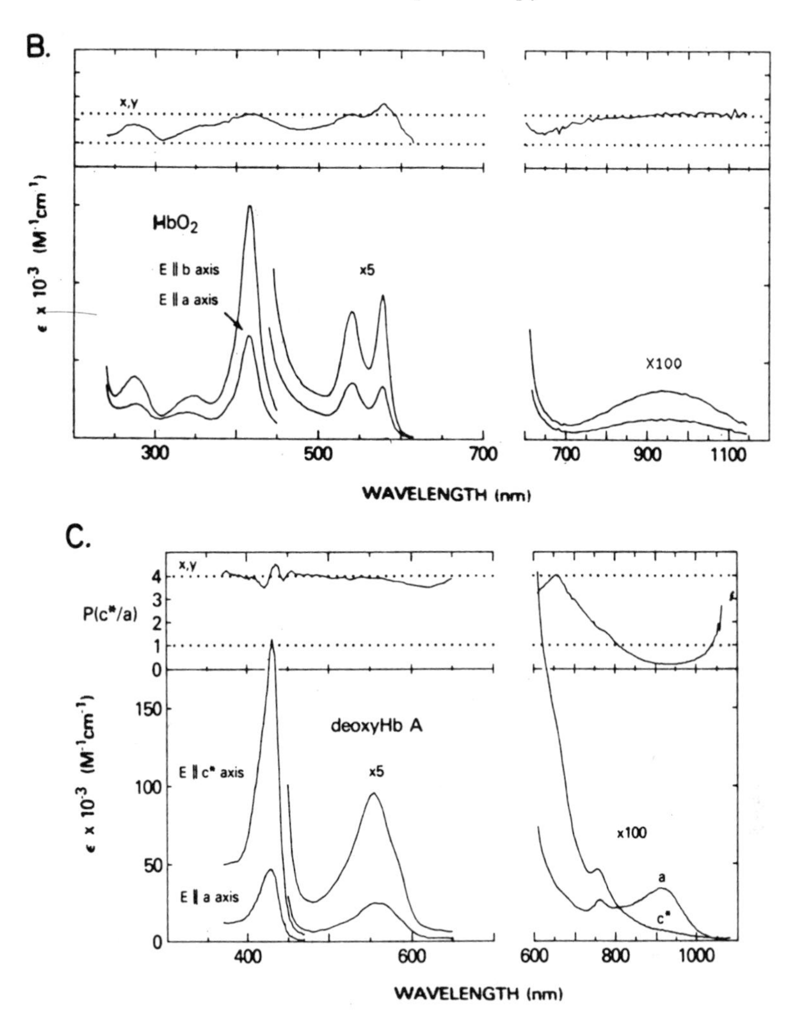

(2) the low-spin type with α- and β-bands in the 530–590 nm region (e.g. alkaline methaemoglobin and the cyanide complex).

The comparatively stable cyanide complex can also be used in the quantitative spectrophotometric assay of haemoglobin and its extinction coefficient is listed in Table 7. Although such ferric species only occur *in vivo* in pathological or other abnormal states, spectra of these types occurring naturally may be diagnostic of the occurrence of peroxidases or catalases, which are catalytically functional in the ferric state.

When polarized light is used, the haemoglobin absorption spectra are seen to be a function of molecular orientation. As Figs. 17B and 17C show, measured

Table 7 Extinction coefficients and wavelengths of maximal absorbance for haemoglobin and its derivatives

Derivative	α-peak		β-peak		γ-peak (Soret)	
	λ_{max} (nm)	E(mM^{-1}cm^{-1})	λ_{max} (nm)	E(mM^{-1}cm^{-1})	λ_{max} (nm)	E(mM^{-1}cm^{-1})
Deoxy Hb*	–		555	12.5	430	133
Oxy Hb*	576	14.6	541	13.5	415	125
Hb CO*	569	13.4	540	13.4	419	191
Hb NO*	575	13.0	545	12.6	418	130
Met Hb†	634	3.8	500	9.0	408	165
Met Hb CN†	–		543	11.0	419	118
Pyridine haemochrome‡	558	30.6	526.5	17.0	419	157

Refs. * Antonini and Brunori (1971)
† Keilin and Hartree (1951)
‡ Gallagher and Elliott (1965)

extinctions of crystalline protein differ when the light is passed parallel to the different crystal axes (corresponding to irradiation of the faces and the edges of the haem rings). Oxyhaemoglobin (Fig. 17B) shows an approximately 60% smaller absorption on one crystal axis throughout the spectrum from 300 to 1100 nm. Deoxyhaemoglobin shows an even bigger difference when examined along the two axes, and a qualitative difference appears in the near-IR region (only one crystal orientation gives absorbance at 900 nm).

Fig. 18 shows that such effects can also be seen with deoxyhaemoglobin *in vivo* when it aggregates (as HbS) in a sickle cell. The aggregated HbS has a partially crystalline structure that gives rise to optically different crystal axes for each cell. In this case plane polarized light is used to discriminate between absorption parallel to and perpendicular to the long axis of the sickled cell. Optical methods can thus be used to probe molecular orientation as well as chemical structure. Haemoglobin, whose chemical behaviour was one of the first to be examined spectroscopically – by Stokes and by Hoppe-Seyler (Keilin, 1966) – is still a favoured subject for spectroscopic analysis, using the present day instrumental descendants of the microspectroscope and the colorimeter.

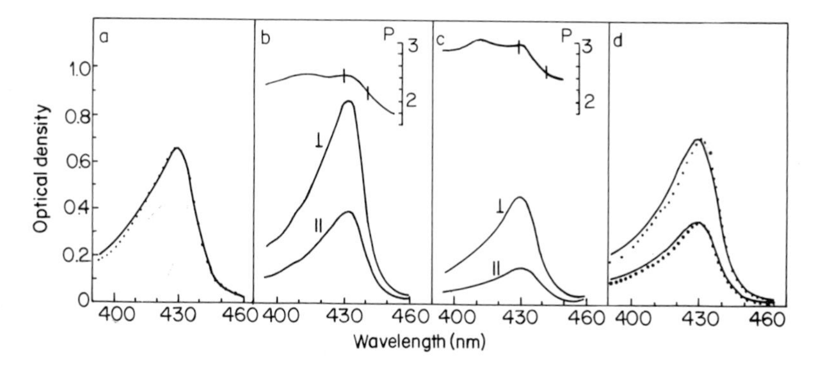

Fig. 18 Polarized absorption spectra in Soret region of deoxygenated human red blood cells. Spectra obtained with (a) microspectrophotometer (sample area 1 μm^2, part of single cell). (a) Normal anaerobic red cell spectrum (●–●) compared with a solution of deoxyhaemoglobin (—). (b) and (c) Polarized spectra of anaerobic sickled cells, light polarized parallel to (II) or perpendicular to (⊥) the long axis of the cell. (d) Isotropic spectra computed from (b) and (c) using Eq. 8, (●–●, O–O) compared with corresponding solutions of deoxyhaemoglobin (—, —).

$$A_{iso} = (2A_{\perp} + A_{\parallel})/3$$

where A_{iso} = isotropic absorbance, $A_{\perp}$ = absorbance measured with light polarized perpendicular to the cell's long axis, $A_{\parallel}$ = absorbance measured with light polarized parallel to the cell's long axis. (Reproduced with permission from Hofrichter *et al.*, 1973)

Acknowledgments

This chapter was completed while on sabbatical leave from Brock University, Canada; I am greatly indebted to Mrs. Stewart of Chelsea College Biochemistry Department for typing, to Chelsea College Audiovisual Unit for drawing several figures, and to the editor, Dr. J. M. Wrigglesworth, for his advice and assistance both academic and technical.

Cited References

Ahmad, I., Cusanovich, M. A., and Tollin, G. (1982) Laser flash photolysis studies of electron transfer between semiquinone and fully reduced free flavins and the cytochrome *c*-cytochrome oxidase complex. *Biochemistry* **21**, 3122–3128

Anderson, J. L., Kuwana, T., and Hartzell, C. R. (1976) Spectroelectrochemical investigations of stoichiometry and oxidation–reduction potentials of cytochrome *c* oxidase components in the presence of carbon monoxide: the 'invisible' copper. *Biochemistry* **15**, 3847–3855

Antonini, E., and Brunori, M. (1971) *Hemoglobin and Myoglobin in their Reactions with Ligands.* North Holland Press, Amsterdam

Beavan, G. H., Holiday, E. R., and Johnson, E. A. (1955) In: *The Nucleic Acids* (eds E. Chargaff and J. N. Davidson), pp. 493–545. Academic Press, London and New York

Brittain, T., Greenwood, C., Springall, J. P., and Thomson, A. J. (1982) The nature of ferrous haem protein complexes prepared by photolysis. *Biochim. Biophys. Acta* **703**, 117–128

Brooks, S. P. J., and Nicholls, P. (1982) Anion and ionic strength effects upon the oxidation of cytochrome *c* by cytochrome *c* oxidase. *Biochim. Biophys. Acta* **680**, 33–43

Brown, S. B. (1980) Ultraviolet and visible spectroscopy. In: *An Introduction to Spectroscopy for Biochemists* (ed. S. B. Brown), Chap. 2, pp. 14–69. Academic Press, London and New York

Chance, B. (1943) The kinetics of the enzyme-substrate compound of peroxidase. *J. Biol. Chem.* **151**, 553–577

Chance, B. (1947) An intermediate compound in the catalase-hydrogen peroxide reaction. *Acta Chem. Scand.* **1**, 236–267

Chance, B., Saronio, C., and Leigh, J. S. (1975) Functional intermediates in the reaction of membrane-bound cytochrome oxidase with oxygen. *J. Biol. Chem.* **250**, 9226–9237

Chance, B., Schoener, B., and De Vault, D. (1965) Reaction-velocity constants for electron transfer and transport reactions. In: *Oxidases and Related Redox Systems II* (eds T. E. King, H. S. Mason and M. Morrison), pp. 907–921. J. Wiley and Sons, New York

Chance, B., and Spencer, E. L. (1959) Stabilization of 'steady states' of cytochromes at liquid nitrogen temperatures. *Disc. Farad. Soc.* **27**, 200–205

Cornish-Bowden, A. J. (1980) *Fundamentals of Enzyme Kinetics.* Butterworth, London

Donovan, J. W. (1973) Ultraviolet difference spectroscopy – new techniques and applications. In: *Methods in Enzymology* (eds C. H. W. Hirs and S. N. Timasheff), Vol. 27, pp. 497–547. Academic Press, New York and London

Eaton, W. A., and Hofrichter, J. (1981) Polarized absorption and linear dichroism spectroscopy of hemoglobin. In: *Methods in Enzymology* (eds E. Antonini, L. Rossi-Bernardi and E. Chiancone), Vol. 76, pp. 175–261. Academic Press, New York and London

Eglinton, D. G., Johnson, M. K., Thomson, A. J., Gooding, P. E., and Greenwood, C. (1980) Near-infrared magnetic and natural circular dichroism of cytochrome *c* oxidase. *Biochem. J.* **191**, 319–331

Erecinska, M., Oshino, R., Oshino, N., and Chance, B. (1973) The *b* cytochrome in succinate-cytochrome *c* reductase from pigeon breast mitochondria. *Arch. Biochem. Biophys.* **157**, 431–445

Estabrook, R. W. (1956) The low temperature spectra of hemoproteins. I. Apparatus and its application to a study of cytochrome *c. J. Biol. Chem.* **223**, 781–794

Gallagher, W. A., and Elliott, W. B. (1965) The formation of pyridine haemochromogen. *Biochem. J.* **97**, 187–193

Guggenheim, E. A. (1926) On the determination of the velocity constant of a unimolecular reaction. *Phil. Mag.* **2**, 538–543

Hartridge, H., and Roughton, F. J. W. (1923) A method of measuring the velocity of very rapid chemical reactions. *Proc. Roy. Soc.* **A104**, 376–394

Hill, B. C. (1982) Kinetic and magnetic studies on mammalian cytochrome oxidase. Ph.D. Thesis, University of East Anglia, Norwich, U.K.

Hiromi, K. (1979) *Kinetics of Fast Enzyme Reactions.* Kodansha, Tokyo, and Wiley, New York

Hofrichter, J., Hendricker, D. G., and Eaton, W. A. (1973) Structure of hemoglobin S fibers: Optical determination of the molecular orientation in sickled erythrocytes. *Proc. Nat. Acad. Sci. USA* **70**, 3604–3608

Jackson, J. B., and Crofts, A. R. (1971) The kinetics of light induced carotenoid changes in *Rhodopseudomonas spheroides* and their relation to electrical field generation across the chromatophore membrane. *Eur. J. Biochem.* **18**, 120–130

Jaffe, H. H., and Orchin, M. (1962) *Theory and Applications of Ultraviolet Spectroscopy.* J. Wiley and Sons, New York and London

Jarrett, H. W., and Penniston, J. T. (1978) Purification of the Ca^{2+}-stimulated ATPase activator from human erythrocytes. *J. Biol. Chem.* **253**, 4576–4682.

Jensen, P., Aasa, R., and Malmström, Bo G. (1981) Electron redistribution in cytochrome *c* oxidase during freezing under turnover conditions. *FEBS Letts.* **125**, 161–164

Keilin, D. (1925) On cytochrome, a respiratory pigment, common to animals, yeast and higher plants. *Proc. Roy. Soc.* **B98**, 312–339

Keilin, D. (1966) *The History of Cell Respiration and Cytochrome.* Cambridge Univ. Press, Cambridge, UK

Keilin, D., and Hartree, E. F. (1945) Properties of azide-catalase. *Biochem. J.* **39**, 148–157

Keilin, D., and Hartree, E. F. (1949) Effect of low temperature on the absorption spectra of haemoproteins: with observations on the absorption spectrum of oxygen. *Nature* **164**, 254–259

Keilin, D., and Hartree, E. F. (1951) Purification of horseradish peroxidase and comparison of its properties with those of catalase and methaemoglobin. *Biochem. J.* **49**, 88–104

Keilin, D., and Hartree, E. F. (1955) Relationship between certain components of the cytochrome system. *Nature* **176**, 200–206

Keilin, D., and Hartree, E. F. (1958) Spectrophotometric study of suspensions of pigmented particles. *Biochim. Biophys. Acta* **27**, 173–184

Kezdy, F. J., Jaz, J., and Bruylants, A. (1958) Cinétique de l'action de l'acide nitreux sur les amides. I. Méthode Générale. *Bull. Soc. Chim. Belg.* **67**, 687–706

Lemberg, M. R., and Legge, J. W. (1949) *Haematin Compounds and Bile Pigments*. Wiley Interscience, New York

Lowry, O. H., and Passonneau, J. V. (1972) *A Flexible System of Enzymatic Analysis*. Academic Press, New York and London

Mahler, H. R., Kline, B., and Mehrotra, B. D. (1964) Some observations on the hypochromism of DNA. *J. Mol. Biol.* **9**, 801–811

Malcolm, A. D. B. (1973) The temperature dependence of the spectroscopic properties of reduced nicotinamide adenine dinucleotide. *Anal. Biochem.* **55**, 278–281

Malkin, R., and Malmström, B. G. (1970) The state and function of copper in biological systems. *Adv. Enzym.* **33**, 177–244

Malviya, A. N., Nicholls, P., and Elliott, W. B. (1980) Observations on the oxidoreduction of the two cytochromes *b* in cytochrome *c*-deficient mitochondria and submitochondrial particles. *Biochim. Biophys. Acta* **589**, 137–149

Minnaert, K. (1961) The kinetics of cytochrome *c* oxidase. I. The system: cytochrome *c*-cytochrome oxidase-oxygen. *Biochim. Biophys. Acta* **50**, 23–34

Netheler, H. G. (1974) Types of photometer. In: *Methods of Enzymatic Analysis* (ed. H. U. Bergmeyer), 2nd English edn, Vol. 1, pp. 184–190. Verlag Chemie Weinheim. Academic Press, New York and London

Nicholls, P. (1982) unpublished observations

Nicholls, P., and Chanady, G. A. (1981) Reactivity of photoreduced cytochrome aa_3 complexes with molecular oxygen. *Biochem. J.* **194**, 713–720

Nicholls, P., and Elliott, W. B. (1974) Cytochromes. In: *Iron in Biochemistry and Medicine* (eds A. Jacobs and M. Worwood), pp. 221–277. Academic Press, New York and London

Nicholls, P., and Kimelberg, H. K. (1968) Cytochromes *a* and a_3: catalytic activity and spectral shifts *in situ* and in solution. *Biochim. Biophys. Acta* **162**, 11–21.

Nicholls, P., and Malviya, A. N. (1968) Inhibition of non-phosphorylating electron transfer by zinc. The problem of delineating interaction sites. *Biochemistry* **7**, 305–310

Nicholls, P., West, J., and Bangham, A. D. (1974) Chlorophyll *b* containing liposomes: Effect of thermal transitions on catalytic and spectral properties. *Biochim. Biophys. Acta* **363**, 190–201

Omura, T., and Sato, R. (1964) The carbon monoxide-binding pigment of liver microsomes. *J. Biol. Chem.* **239**, 2370–2378

Penzer, G. R. (1980) Molecular emission spectroscopy (Fluorescence and phosphorescence). In: *An Introduction to Spectroscopy for Biochemists* (ed. S. B. Brown), pp. 70–114. Academic Press, New York and London

Racker, E. (1976) *A New Look at Mechanisms in Bioenergetics*. Academic Press, New York and London

Rayleigh, Lord (1899) On the light from the sky, its polarization and colour. In: *Scientific Papers* Vol. I, pp. 87–103. Cambridge University Press

Sato, R., and Omura, T. (1978) *Cytochrome P450*. Kodansha, Tokyo, and Academic Press, New York

Sauer, K. (1975) Primary events and the trapping of energy. In: *Bioenergetics of Photosynthesis* (ed. Govindjee), pp. 116–181. Academic Press, New York and London

Sawicki, C. A., and Morris, R. J. (1981) Flash photolysis of hemoglobin. In: *Methods in Enzymology* (eds E. Antonini, L. Rossi-Bernardi and E. Chiancone), Vol. 76, pp. 667–681. Academic Press, New York and London

Shipp, W. S. (1971) Multiple *b* cytochromes in rat and pigeon mitochondria detected by finite difference analysis of absorption spectra. *Biochem. Biophys. Res. Commun.* **45**, 1437–1443
Stern, K. G. (1936) On the mechanism of enzyme action. A study of the decomposition of monoethyl hydrogen peroxide by catalase and of an intermediate enzyme-substrate compound. *J. Biol. Chem.* **114**, 473–494
Tyndall, J. (1863) *Six Lectures on Light*. Longmans, London
van Buuren, K. J. H., Nicholls, P., and van Gelder, B. F. (1972) Biochemical and biophysical studies on cytochrome aa_3. VI. Reaction of cyanide with oxidized and reduced enzyme. *Biochim. Biophys. Acta* **256**, 258–276
Waterman, M. (1978) Spectral characterization of human hemoglobin and its derivatives. In: *Methods in Enzymology* (eds S. Fleischer and L. Packer), Vol. 52, pp. 456–463. Academic Press, New York and London
Williams, J. (1964) A method for the simultaneous quantitative estimation of cytochromes a, b, c_1 and c in mitochondria. *Arch. Biochem. Biophys.* **107**, 537–543
Williams, W. P. (1983) Fluorescence. Chapter 2, this volume

Books Recommended for Further Reading

(a) General

Beavan, G. H., Johnson, E. A., Willis, H. A., and Miller, R. G. J. (1961) *Molecular Spectroscopy*. Macmillan and Co., London
Brown, S. B. (ed.) (1980) *An Introduction to Spectroscopy for Biochemists.* Academic Press, London and New York
Freifelder, D. (1982) *Physical Biochemistry* 2nd edn. (esp. Chapter 14, Absorption Spectroscopy). Freeman and Co., San Francisco

(b) Haemoglobins

Antonini, E., Rossi-Bernardi, L. and Chiancone, E. (eds) (1981) *Methods in Enzymology*, Vol. 76. Academic Press, New York and London

(c) Stopped Flow and Other Rapid Kinetic Methods

Hiromi, K. (1979) *Kinetics of Fast Enzyme Reactions.* Kodansha, Tokyo, and Wiley, New York

Biochemical Research Techniques
Edited by J. M. Wrigglesworth

2
Fluorescence

W. PATRICK WILLIAMS

Department of Biophysics and Bioengineering, Chelsea College, University of London, UK

2.1 Introduction

The measurement of fluorescence parameters has been recognized as a major tool in biological studies for several decades. Widespread application of fluorescence techniques has, however, been hampered by the limited availability of reliable, non-expensive fluorimeters and spectrofluorimeters. The development of new experimental methods involving the use of fluorescent probe and labelling techniques, together with the increased availability of suitable measuring equipment, has led to a rapid increase in the use of fluorescence techniques in many areas of research over recent years.

This chapter is intended to provide a brief introduction into the principles of fluorescence, simple fluorescence instrumentation and some of the current applications of fluorescence measurements. Emphasis is placed on the fundamental principles of the subject and the avoidance of the many potential pitfalls associated with fluorescence measurements. A list of reviews and more detailed texts is provided at the end of the chapter for those readers interested in a more comprehensive treatment of particular points.

2.2 Fluorescence and Fluorescence Parameters

2.2.1 Basic Theory

The absorption, or emission, of light by a given molecule is associated with a transition between two electronic states. Each electronic state is associated with corresponding sets of vibrational and rotational energy levels. The relationship between these different levels is best illustrated in terms of a potential energy diagram. Although such diagrams can only be drawn conveniently for diatomic molecules, the fact that they provide a clear picture of the transitions of interest in fluorescence studies (and of the limitations imposed on such transitions by molecular geometry) means that they provide a useful starting place for discussion of fluorescence phenomena.

A schematic version of a typical potential energy diagram for a diatomic molecule is shown in Fig. 1. The diagram shows the potential energy wells, and associated vibrational levels, of the ground (S_0) and first excited state (S_1) of the singlet manifold of the molecule together with the first excited state (T_1) of the triplet manifold. The terms singlet and triplet refer to the multiplicity of the

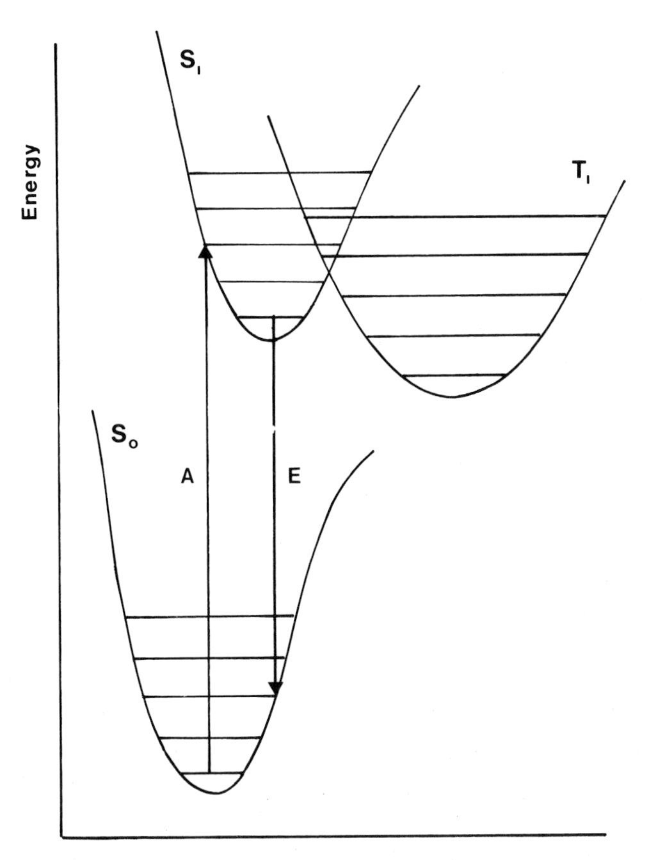

Fig. 1 Potential-energy diagram for a diatomic molecule showing the electronic and vibrational levels of the ground state and the first excited states of the singlet (S) and triplet (T) manifolds. The vertical lines correspond to transitions associated with the absorption of a photon (A) and the subsequent emission of a photon (E) in the form of fluorescence

electron spin states associated with the molecule. The electrons of a given molecule can normally exist in one of two possible spin states characterized by a spin quantum number, m_s, where $m_s = \pm 1/2$. In most molecules, the electrons are arranged in different molecular orbitals so that they are spin paired and, S, the total spin quantum number of the molecule (where $S = \Sigma m_s$), is equal to zero. If, as is sometimes the case, the molecule contains two degenerate orbitals each containing a single electron both of which have parallel spins, $S = 1$.

The terms singlet and triplet arise from the fact that the spinning electron has an associated magnetic moment. Molecules in the singlet state ($S = 0$) show no preferred orientation in an external magnetic field and thus have a single energy level whilst those in the triplet state ($S = 1$) can take up any one of three possible orientations. Almost all molecules form singlet ground states. The only important exception is molecular oxygen which exists as a triplet in the ground state.

The probability of transition between any two electronic states is governed by quantum mechanical selection rules. In order for light to interact with matter, the interaction must be accompanied by a change in the distribution of electrons. The selection rules state that such transitions must involve appreciable dipole moments of transition. Transitions within the singlet, or triplet, manifold normally involve large transition moments but those between manifolds do not. In terms of quantum mechanics, spin–reversal is a 'forbidden' process. This does not mean that direct excitation from a singlet ground state to a triplet excited state cannot occur. Such transitions are, however, extremely rare accounting for perhaps one in 10^6 excitations. Entry to the triplet manifold, if it occurs, normally takes place via excited singlet states.

When a molecule absorbs a photon, one of its electrons is promoted to a higher electronic state. This process is very rapid, taking place within $\sim 10^{-15}$ sec (i.e. on a time scale similar to that of the oscillatory period of the incident radiation). As nuclear vibration periods are typically of the order of 5×10^{-13} sec, the nuclear coordinates of the molecule are effectively constant during excitation. A direct consequence of this is that the transition between the ground and excited states corresponds to a vertical line on potential energy diagrams of the type shown in Fig. 1. This is referred to as the Franck–Condon principle and the molecule immediately after excitation is said to be in its Franck–Condon state.

If the shape of the potential energy wells of S_0 and S_1 were to be identical, the transition between the two states would involve only a change in the electronic energy of the molecule. The molecular orbital configurations of the ground and excited states, however, normally show appreciable differences. This, as illustrated in Fig. 1, is reflected in differences both in the positions of the potential energy minima (corresponding to the equilibrium internuclear distances between the atoms of the molecule) and the shapes of the curves themselves (reflecting differences in the force constants associated with the vibrational motion of the atoms). The internuclear distances of the excited states are usually rather greater than those of the corresponding ground state. As a consequence, the transition of a molecule from its ground state to an excited state results in a transfer to a higher vibrational level than that occupied in the ground state.

The fraction of the total population of molecules existing in a particular vibrational level under equilibrium conditions can be predicted using the Boltzmann distribution. At room temperature, nearly all the molecules will exist in the lowest vibrational level of the ground state. Immediately following excitation, the excited molecule normally possesses considerable excess vibrational energy. This excess energy is rapidly dissipated as heat by the process of *vibrational relaxation* within a time scale of 10^{-11}–10^{-12} sec. At the same time, the molecule will attain an equilibrium nuclear geometry dictated by the excited state electron distribution and any rearrangements of the solvation envelope required to accommodate this distribution will take place.

Following the completion of these relaxation processes, the molecule normally remains in its equilibrium excited state for a period of about 10^{-9} sec. The molecule then usually relaxes to its ground state either by means of a radiationless transition, in which the excess electronic energy is dissipated as heat, or by a radiative transition involving the emission of a photon. The radiationless process is termed *internal conversion* and the radiative process *fluorescence*. Excitation to higher electronic levels of the singlet manifold is almost invariably followed by rapid internal conversion to the first excited state. Fluorescence emission direct from higher excited states is extremely rare.

An alternative pathway to the ground state exists via the triplet state. In some molecular species, spin reversal in the excited state can effectively compete with radiative and radiationless decay direct to the ground state. As the electronic energy associated with the triplet state is almost always lower than that of the corresponding singlet state, electrons entering the triplet state by *intersystem crossing* tend to possess considerable excess vibrational energy. This, as in the singlet state, is rapidly dissipated as heat by vibrational relaxation leaving the molecule in the lowest vibrational energy level of the triplet manifold.

In order to return to the ground state, the electron must again undergo spin reversal. The molecule is thus trapped in the triplet state until it can either gain sufficient kinetic energy from its surroundings by virtue of collisions to reverse the process of intersystem crossing or return directly to the ground state by a radiative or radiationless transition involving spin reversal. The former process is energetically very unfavourable but does occur occasionally giving rise to the phenomenon of *delayed fluorescence*. The normal mode of decay from the triplet state is either by a radiative transition, giving rise to *phosphorescence*, or by a radiationless process of internal conversion. In either case, these processes are quantum mechanically forbidden and take place on a millisecond to second time scale as opposed to the nanosecond time scale associated with fluorescence.

Most molecules used in fluorescence studies are too complicated to be represented using potential energy diagrams. Jablonski energy level diagrams provide a much more convenient method for showing the energy changes associated with the transitions of interest in fluorescence studies in such molecules. A formalized energy level diagram showing the various processes described above is set out in Fig. 2. Such diagrams lack the detailed spatial

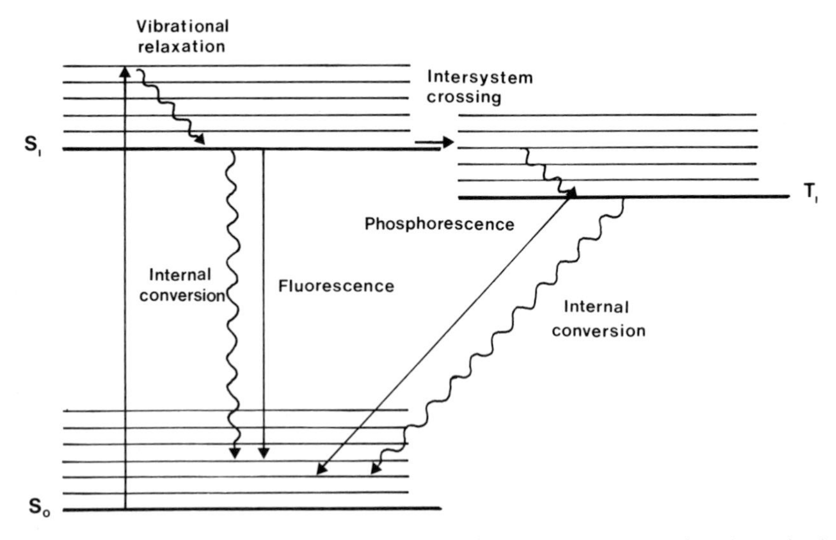

Fig. 2 Jablonski energy level diagram showing the various competing de-excitation processes available to molecules in the excited state

information associated with potential energy diagrams but this is more than compensated for by the fact that they provide a much easier appreciation of the energetics of the processes involved.

2.2.2 Fluorescence Parameters

2.2.2.1 Fluorescence Emission Spectrum

The emission spectrum represents the variation of fluorescence intensity with emission wavelength. It lies to the longer wavelength (lower energy) side of the longest wavelength band of the absorption spectrum. The displacement of the peak of the emission band with respect to the peak of this absorption band is referred to as the *Stokes shift.* This shift is a direct reflection of the loss of excitation energy associated with vibrational relaxation. Its magnitude depends on the differences in geometry of the excited state with respect to the ground state. In many cases, the fluorescence emission spectrum, after due correction for instrument response, is an approximate mirror image of the lowest energy absorption band. This symmetry arises from the fact that transitions making up the fluorescence emission spectrum, after allowance for vibrational relaxation, are in effect the reverse of absorption transitions between the lowest vibrational level of the ground state and the different vibrational levels of the first excited state.

2.2.2.2 *Fluorescence Excitation Spectrum*

The excitation spectrum represents the variation of fluorescence intensity with the wavelength of the exciting light. The fact that the excited molecules normally relax to the lowest vibrational level of the first excited state prior to the emission of fluorescence means that all absorbed photons are equally likely to give rise to emission. This in turn means that in the case of a pure solution containing a single fluorescent species, the excitation spectrum should be directly equivalent to the absorption spectrum. It should, however, be noted that the measured excitation spectrum must be corrected for equal numbers of incident photons at all excitation wavelengths for this to be true.

2.2.2.3 *Quantum Yield of Fluorescence*

The quantum yield of fluorescence, Φ, is defined as:

$$\Phi = \frac{\text{number of quanta emitted as fluorescence}}{\text{total number of quanta absorbed}} \tag{1}$$

Its value depends on the relative rates of the various de-excitation processes available to the excited molecule. These processes include the radiative and radiationless processes outlined in the previous section together with quenching of fluorescence by collision with other molecules or interaction with the solvent.

Measurements of the absolute quantum yield of fluorescence are technically extremely difficult as they involve the summation of fluorescence over all wavelengths and emission angles. For most purposes comparative measurements made under specified measuring conditions are quite adequate. These measurements, which are usually expressed in relative units, are normally referred to as *fluorescence yield* values.

2.2.2.4 *Fluorescence Lifetime*

The lifetime of fluorescence, τ, is defined as the time required for the fluorescence intensity to fall to $^1/_e$ of its value when the exciting light is initially turned off. Another useful quantity is the *natural (or intrinsic) lifetime*, τ_0, which is defined as the anticipated lifetime if the only de-excitation process is fluorescence decay. It is related to the measured lifetime by the relationship:

$$\tau = \Phi\tau_0 \tag{2}$$

2.2.2.5 *Fluorescence Polarization*

The absorption of a photon is invariably accompanied by a change in the spatial distribution of the electrons. This change, as we have already noted, is associated with a transition moment. This moment, often referred to as the transition dipole of the molecule, is a vectorial quantity and efficient absorption takes place only when it is aligned with the electric field oscillation of the exciting light. If a population of randomly oriented molecules are illuminated by a beam of plane polarized light, those molecules with transition dipoles most closely aligned with the oscillating electric field will be preferentially excited. Fluorescence emission is associated with a reversal of the dipole moment change accompanying absorption. Changes in molecular orientation during the lifetime of the excited state can be readily detected by fluorescence polarization measurements. The two most commonly used parameters are the *fluorescence polarization*, P, and *fluorescence polarization anisotropy*, r. Their values are defined by the relationships:

$$P = \frac{I_{\parallel} - I_{\perp}}{I_{\parallel} + I_{\perp}} \; ; r = \frac{I_{\parallel} - I_{\perp}}{I_{\parallel} + 2I_{\perp}} \tag{3}$$

where $I_{\parallel}$ and $I_{\perp}$ are the fluorescence emission intensities measured parallel and perpendicular to the plane of the polarized exciting beam.

2.2.2.6 Energy Transfer and Sensitised Fluorescence

Excitation energy can, under appropriate circumstances, be transferred from an excited molecule to a neighbouring non-excited molecule of the same, or different, species. The process involves a direct transfer of energy by a coupling of the dipole fields of the two molecules and should be clearly distinguished from the trivial process of emission and reabsorption. The probability of transfer, P_t, by this process is given by the expression:

$$P_t = \frac{KA}{\tau_0 R^6} \int_{\nu}^{\varepsilon} \frac{\varepsilon(\nu) f(\nu)}{\nu^4} \, . \, d\nu \tag{4}$$

where $\varepsilon(\nu)$ represents the spectral distribution of the extinction coefficient of the acceptor molecule, $f(\nu)$ is the spectral distribution of emission of the donor molecule, R is the distance between the donor and acceptor, K is a function of their mutual orientation and A is a characteristic of the environment dependent on its refractive index.

The physical significance of the integral in this expression is that it provides a measure of the area of overlap between the emission band of the donor and the absorption band of the acceptor (see Fig. 3). It gives a measure of the probability that the two molecules share transitions of equal energy that can be coupled in the transfer process. The fact that the transfer efficiency is inversely proportional to R^6 means that the process is extremely short-range. Nevertheless high-efficiency transfer is possible over separations of up to 20–30 Å. Rates of energy transfer can be determined from measurements of the quenching of the fluorescence of the donor species by the acceptor. If the acceptor is fluorescent,

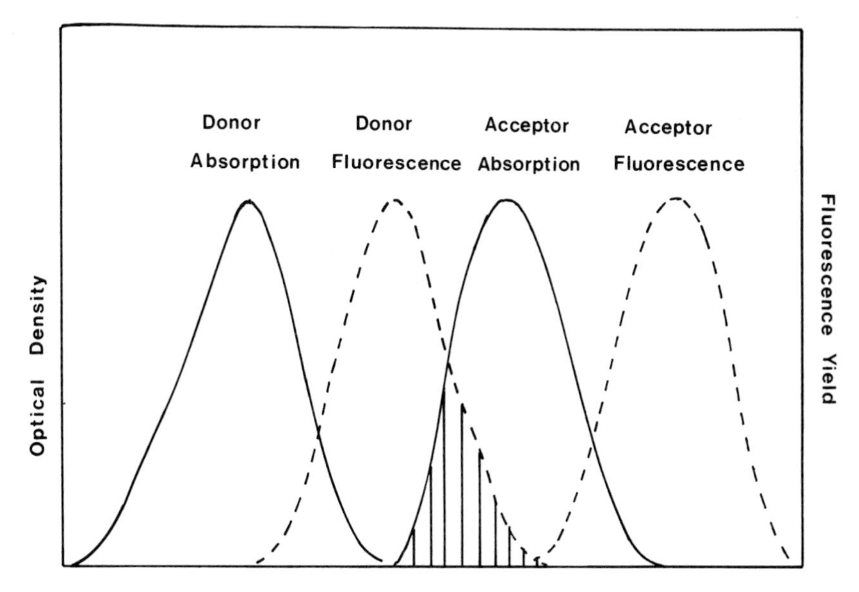

Fig. 3 Diagram illustrating the relationship between the absorption and emission spectra required for resonance energy transfer in a hypothetical donor–acceptor system. The hatched region shows the overlap area corresponding to the existence of common energy levels in the emission spectrum of the donor and the absorption spectrum of the acceptor

transfer rates can also be estimated from measurements of *sensitized fluorescence* emission from the acceptor.

2.3 Instrumentation

Most applications of fluorescence techniques in the biological sciences involve the measurement of steady state parameters using standard fluorimeters and spectrofluorimeters. Attention is therefore focused here on this type of instrumentation. The use of fluorescence lifetime measurements, although much less widespread, is slowly increasing and a brief introduction to such techniques is also provided. Brief accounts of suitable instrumentation for phosphorescence and bioluminescence studies are also provided.

2.3.1 Steady State Instruments

The basic elements of any steady state fluorimetry system consist of a light source, a means of isolating monochromatic (or narrow band) light for sample excitation, a sample cuvette, a means of isolating the fluorescence emission from scattered exciting light, a photodetector and a recorder system. Commercially available instruments range from those using simple filter systems (usually referred to as fluorimeters), through attachments to absorption spectrophotometers to dedicated spectrofluorimeters of varying complexity. The reliability and sensitivity of such instruments, it is important to appreciate, runs through a similarly wide range.

A block diagram showing the main features of a modern high-performance spectrofluorimeter is set out in Fig. 4. The main differences between this system and the basic system described above are the incorporation of a light-modulation system and lock-in-amplifier to allow the instrument to be operated in a modulated (a.c.) mode and a ratio-recording facility to improve signal stability.

One of the main problems associated with the design of any fluorimetry system is the elimination of stray light. Stray light is light that enters, and passes through, the optical system giving rise to spurious signals at the photodetector. It can enter the system at many points. Poor quality monochromators, for example often allow the passage of appreciable amounts of light of all wavelengths. Ill-fitting lids and casings can also allow light leakage into the sample compartments or other parts of the system. Whilst such problems can

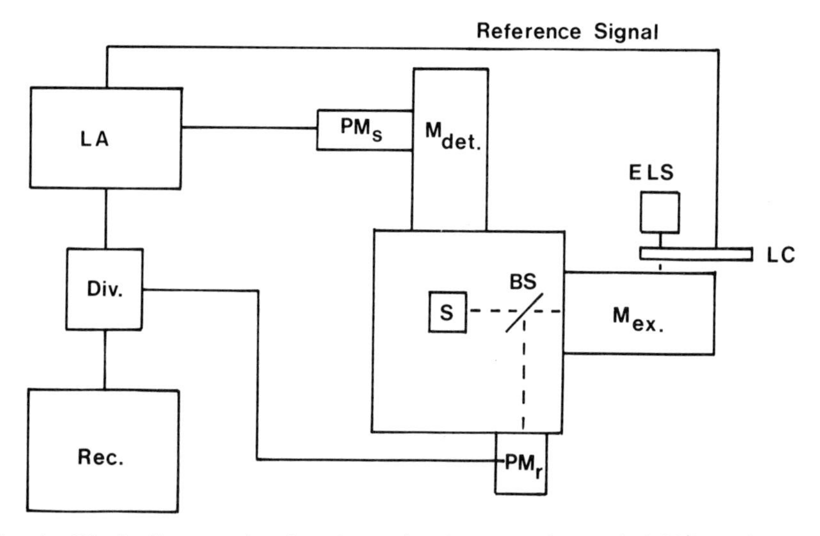

Fig. 4 Block diagram showing the main elements of a typical high-performance spectrofluorimeter. Light from the excitation light source (ELS) is modulated by a light-chopper (LC) and passed through a monochromator (M_{ex}). Emission from the sample (S) is collected by a second monochromator (M_{det}) and monitored by a photomultiplier (PM_S). Part of the excitation beam is reflected from a beam splitter (BS) and its intensity monitored by a reference photomultiplier (PM_r). The amplitude of the modulated fluorescence signal, isolated using the lock-in-amplifier (LA), is compared to the amplitude of the signal from (PM_r) in a divider circuit (Div) and the corrected signal fed to a recorder (Rec.)

be minimized by the use of high quality monochromators and good instrumental design, the use of a.c. modulation techniques has proved particularly useful in reducing stray-light signals.

Instruments operating in the a.c. mode use a modulated excitation beam rather than the usual continuous beam. The excitation beam is modulated by passing it through a light-chopper consisting of a rotating, slotted disc; the frequency of rotation of the disc determining the modulation frequency of the exciting light and hence of fluorescence emission. A lock-in-amplifier system is then used to isolate the modulated component of the photodetector signal. Non-modulated components arising from stray-light signals are rejected. Stray- light components modulated together with the initial excitation beam cannot be eliminated in this way. These, however, usually represent a negligible fraction of the total stray light.

Ratio-recording is used to minimize fluctuations in fluorescence output associated with random fluctuations in the intensity of the excitation beam. Most measurements of biological interest involve the use of excitation wavelengths between about 250 and 450 nm. In order to obtain sufficiently high intensities in this wavelength region, it is normally necessary to use xenon or mercury arc sources. Such sources require well-stabilized power supplies and the relatively poor long-term stability of some commercially available spectrofluorimeters can be directly attributed to the use of inadequately stabilized power supplies. Another major problem associated with the use of electric-arc sources is the tendency for the path of the arc to 'wander' as the electrode surfaces wear down. Minor changes in the path of the arc can lead to appreciable changes in the position of its image at the input slits of the excitation monochromator and hence to the intensity of light transmitted to the sample.

In ratio-recording systems a small fraction of the light emerging from the excitation monochromator is diverted to a reference photomultiplier by reflection from the surface of a partially silvered mirror (beam splitter). This photo multiplier is used to monitor the intensity of the excited light. The final fluorescence signal is then taken as the ratio of the outputs of the photomultiplier monitoring the fluorescence signal and this reference photomultiplier. The divider circuit used to compare the two signals compensates for any fluctuation in excitation intensity.

Most commercial spectrofluorimeters produce 'uncorrected' fluorescence emission and excitation spectra. In order to convert these to 'corrected' spectra, it is necessary to make allowances for the wavelength dependence of the detection system in the case of emission spectra, and for equal numbers of incident photons in the case of excitation spectra. A number of systems, based on the use of dye standards as 'quantum counters', are available as optional attachments on more sophisticated instruments. These until recently were largely analogue devices based on the use of electronically matched reference photomultipliers. The use of minicomputers and microprocessors for the preprocessing of fluorescence data is, however, rapidly expanding and most commercial correction devices are now microprocessor controlled.

2.3.2 Fluorescence Lifetime Measurements

Although the techniques used in the measurement of fluorescence lifetimes are well-established, the expense and complexity of the instrumentation involved has meant that its use has been restricted to a few specialized laboratories. Only a brief introduction to the topic is provided here. Further details can be found in the references given at the end of the chapter.

Instruments for the measurement of fluorescence lifetime divide into two main categories; phase-shift and repetitive-flash instruments. The basic elements of these two groups of instruments are illustrated diagrammatically in Fig. 5.

Phase-shift instruments measure the phase difference between the incident beam and the emitted light. This is related to the fluorescence lifetime of the emitting species by the expression:

$$\tan \varphi = \omega\tau \tag{5}$$

where φ is the phase difference between the beams and ω their modulation frequency. This relationship is based on the assumption that there is a single fluorescing species and that it decays exponentially with time. The method is thus unsuitable for the analysis of system showing complex decay patterns.

The repetitive-flash technique involves the excitation of fluorescence by a short duration flash, or laser pulse, and the direct measurement of the time dependence of the decay of the emitted fluorescence. The intensity of the individual flashes is low and the results obtained from a large number of repetitive flashes are averaged to increase signal-to-noise ratio. The main difficulties encountered in such measurements are associated with the production of sufficiently short duration flashes. As the fluorescence decay is in the nanosecond range the flash should ideally be at least an order of magnitude shorter. This, in practice, is difficult to achieve and the emission signal contains contributions from molecules excited throughout the whole flash period. It is usually necessary to employ complex deconvolution procedures to extract the lifetime data from the measured signal. Nevertheless, repetitive-flash methods, particularly in combination with single-photon counting systems, have largely taken over from phase-shift methods.

2.3.3 Phosphorescence and Bioluminescence Measurements

Phosphorescence can be measured using exactly the same equipment as used for fluorescence measurements. The only difficulties arise when it is desired to

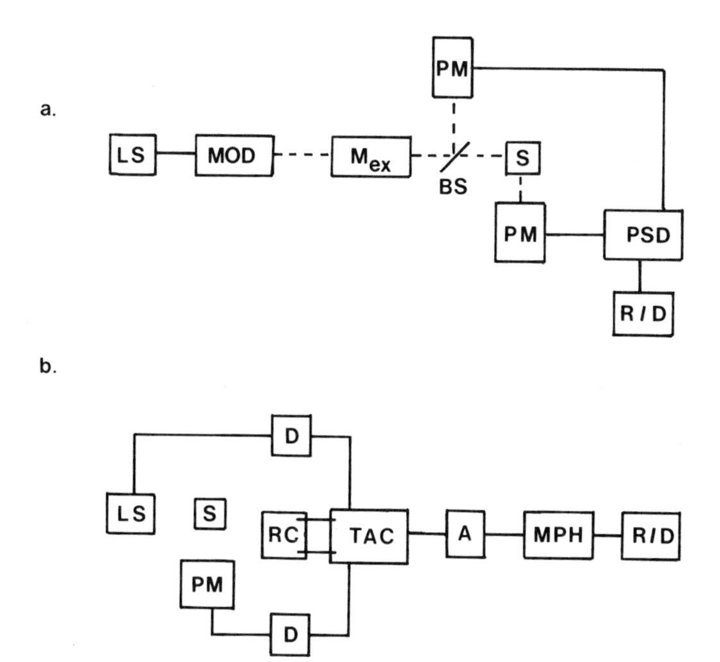

Fig. 5 Block diagrams showing the main elements of (a) a phase-modulation, and (b) a multiple-pulse system for the measurement of fluorescence lifetimes. In the phase-modulation system, light from the light source (LS) (xenon arc) is modulated using a piezoelectric crystal (MOD) and the light passed through a monochromator (M_{ex}). A fraction of the excitation light is diverted using a beam splitter (BS). The phase of the exciting light and the light emitted by the sample (S) are monitored by photomultipliers (PM) and compared using a phase-sensitive detector (PSD). The resulting signals are fed to a recorder/display system (R/D). In the multiple-pulse system, the light source gives rise to a series of short, intense light pulses. Pulse discriminators (D) are used to identify excitation pulses and pulses in the photomultiplier output and their numbers are compared in the ratio-recorder (RC). The response of the system is adjusted so that not more than one photon event is detected per excitation pulse. A time-amplitude converter (TAC) is used to measure the time lapse between the excitation pulse and photon emission. Its output passes through an amplifier (A) to a multichannel pulse-height analyser (MPH). Results are accumulated over a large number of pulses and then fed to recorder/display systems

measure phosphorescence against a background of fluorescence emitted at the same wavelength. Normal spectrofluorimeters do not distinguish between the two types of emission. Under these circumstances, it is necessary to use either a phosphoroscope or a phosphorescence attachment that can be fitted to a spectrofluorimeter. In either case, the general principle of the measurement is the same; the excitation and detection processes are temporally separated by some form of mechanical shutter.

As fluorescence decays on a nanosecond time scale whilst phosphorescence decays on a millisecond time scale, temporal separation of the two types of emission is quite straightforward. A number of different systems are available. One of the simplest, involving the use of a pair of rotating discs is illustrated in Fig. 6. Phosphorescence measurements are made during the period in which the excitation beam is occluded. With appropriate modifications of the discs and recording system phosphorence and fluorescence can be measured simultaneously on the same instrument.

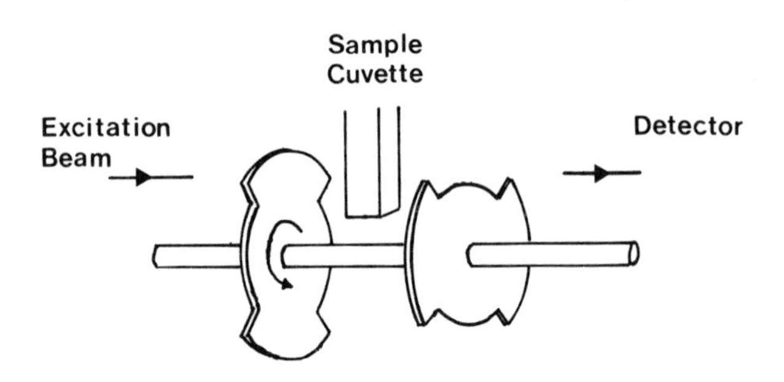

Fig. 6 Diagrams showing the general principle of the Becquerel phosphoroscope

The term bioluminescence refers to light emission from excited states generated by biochemical reactions rather than by the absorption of light energy. Measurements of bioluminescence, like phosphorescence, can be performed using either standard fluorescence equipment (appropriately modified to allow injection of the reactants, stirring, etc.) or by use of specially designed instruments. A number of simple luminometer systems are commercially available that are specifically designed for chemiluminescence and bioluminescence measurement. A block diagram of such an instrument is shown in Fig. 7. Light emitted by the sample is collected by a reflector system. This light falls on a photomultiplier, the output of which is amplified and passed on to a recorder or digital printer as required.

2.4 Measurement Considerations

2.4.1 Choice of Instrumentation

The choice of suitable instrument is largely dictated by the type of measurement, or measurements, contemplated by the individual worker. Some simple routine assays can be accurately and reliably carried out using relatively cheap, unsophisticated filter fluorimeters. Other applications are far more demanding. The important criteria in assessing the suitability of a given instrument for any particular application can be summarized as the three 'Ss' – spectral purity, stability and sensitivity.

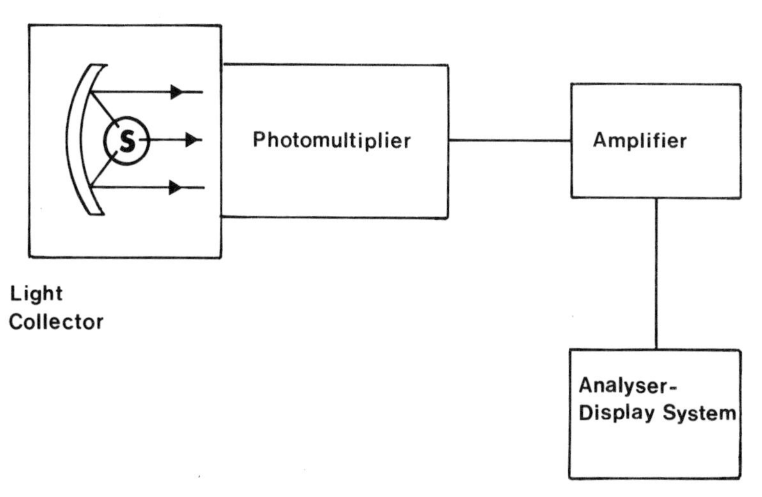

Fig. 7 Block diagram showing the general layout of a typical luminometer system of the type used in bioluminescent assays of ATP

Spectral purity refers to the necessity of being sure that the measured emission signal is a true reflection of fluorescence and contains no significant contributions from stray light or scattered excitation light. A good test for spectral purity is to measure the emission spectrum of a known compound and to compare it, after appropriate correction for the wavelength sensitivity of the detection system, with published spectra. One of the major disadvantages of filter fluorimeter systems is the fact that simple checks of this type cannot be made. When using filter instruments, it is essential to perform rigorous control measurements using different filter combinations and reference blanks showing similar scattering properties to the samples of interest in order to check that no extraneous light reaches the photodetector.

The importance of long-term stability cannot be overstressed. Many commercial instruments, particularly some of the cheaper instruments using xenon or mercury arc sources, show very poor signal stability. It is important to be sure that no appreciable signal drift occurs in the time scale of the planned measurements. This is easily tested by switching the instrument to an appropriate sensitivity at the desired wavelength settings and monitoring the output of a test sample over a suitable period.

Sensitivity refers to the ability of the instrument to provide an accurate, stable measurement of low concentrations of the fluorescing species of interest. This is primarily determined by the stray-light rejection of the instrument and the intensity and stability of the excitation source. These must all be high in order to ensure high sensitivity.

Unfortunately, there are no generally agreed standards for the comparison of the sensitivity of different spectrofluorimeters. It is, in principle, easy to prepare fixed concentrations of chemicals of known quantum yield and to compare the emission spectra produced by different instruments under defined conditions. This, however, has not been generally attempted. A very simple testing technique, that is employed by at least one manufacturer, is to use the Raman scattering peak of distilled water as a standard (see Section 2.4.4). It must be emphasized that absolute sensitivity is not necessarily the most important criterion in the choice of a suitable instrument. High sensitivity is expensive to achieve and sensitivity requirements vary from application to application. The requirements for spectral purity and signal stability are, however, basic to all fluorescence measurements.

2.4.2 Choice of Excitation and Emission Wavelengths

The general principle involved in the choice of excitation and emission wavelengths is that the chosen wavelength should give the maximum signal whilst avoiding problems associated with the light-scattering, internal-filter and fluorescence reabsorption artefacts described below. If the application in hand simply involves measurements of changes in fluorescence efficiency at a given wavelength, potential scattering problems can usually be minimized by the use of appropriate cut-off filters in the excitation and/or detection pathways. Filters may also be required to eliminate any second-order diffraction maxima present in the exciting light. If more than one fluorescing species is present, care has also to be taken to choose excitation and/or emission wavelengths that avoid spectral overlap of the different species. If this is not possible, due allowance for such overlap must be made in calculating fluorescence yields.

2.4.3 Choice of Slit Width

The size of the fluorescence signal can be increased either by increasing the slit widths of the excitation and/or detection monochromators or by increased signal amplification. The strategy adopted depends on the type of measurement in hand. Assuming that the incident light fully fills the entrance slits, doubling the slit widths of a monochromator will normally lead to a four-fold increase in the amount of light transmitted. This is achieved without the concomitant increase in noise level associated with signal amplification. It must be stressed, however, that if the emission signal contains contributions from broad-band stray or scattered light that these are likely to be proportionately increased by the use of wider slits. Wider slit widths can also give rise to problems if overlapping emission bands are present.

An important factor that must be borne in mind if the shape of the emission, or excitation, band is of particular interest is that the band-pass of the appropriate monochromator should normally be less than one-fifth of the half-band width of the spectral feature of interest. The use of overly large slit widths can lead to a degradation of the spectrum and a loss of important spectral features.

2.4.4 Scattered Light

Fluorescence measurements on samples of biological interest often involve the use of highly turbid samples. Light-scattering artefacts can be a major problem in such measurements (see also Chapter 1). When light passes through a transparent medium a fraction of it is inevitably scattered. This scattering is of two types; Rayleigh scattering in which the light is elastically scattered and Raman scattering in which it is inelastically scattered.

Rayleigh scattering is by far the more important. It gives rise to a single scattering peak centred at the wavelength of the incident beam. Raman scattering corresponds to scattered light which has lost (or gained) energy by the transfer of one, or more, vibrational quanta between the light and scattering centre. In practice as most of the scattering molecules are in their vibrational ground states, and the energetics of the transfer process favour the transfer of a single vibrational quantum, only the first low energy (long wavelength) Raman peak is of appreciable intensity. The relative sizes of the Rayleigh and Raman scattering peaks of pure distilled water are shown in Fig. 8. The presence in the sample of protein, or cellular material usually leads to a marked broadening of the Rayleigh scatter peak swamping any effects due to Raman scattering. Under these conditions, the possibility of complications due to Raman scattering can normally be ignored.

The two main problems associated with light scattering are the recognition of its presence and its subsequent elimination. The presence of light scattering can easily be detected, and corrected for, by use of a reference that differs from the sample only in lacking the fluorescing species. Unfortunately, such reference samples are sometimes not available. If the emission spectrum of the fluorescing species is known, light scattering can usually be recognized by comparing the known and measured emission spectra. Even if the true spectrum is not known, comparison of emission spectra measured at different exciting wavelengths will often reveal any distortions due to scattered light. Similar comparisons of excitation spectra can also be made. If none of these methods is applicable and light scattering is suspected, it is often useful to make up a reference of similar turbidity to the sample using a convenient colloidal suspension (e.g. diluted milk) and to check that for light scattering.

Several strategies are available for the minimization of light scattering effects. The first, and simplest, is to increase the separation of the excitation and detection wavelengths. As an added precaution, cut-off filters can be placed in the excitation and detection pathways. Attention should also be paid to the slit widths of the excitation and detection monochromators. Slit widths should be kept as narrow as is commensurate with a reasonable signal-to-noise

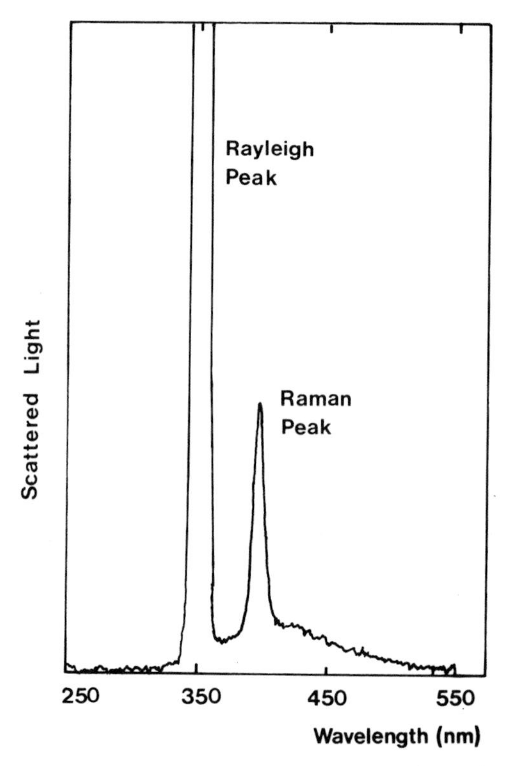

Fig. 8 Rayleigh and Raman light-scattering peaks of distilled water measured using a Perkin-Elmer MPF-44A Spectrofluorimeter. The excitation wavelength was 350 nm and the excitation and emission slits were both set at 5 nm half-band widths

ratio. This will reduce the size of the fluorescence signal but such reductions are worthwhile if the light scattering can be eliminated or preferentially reduced.

An alternative approach is to modify the sample holder. In most fluorimeters, fluorescence is detected at right angles to the excitation beam. This means that the detection monochromator will receive light from an extended source that is effectively the width of the sample cuvette (Fig. 9a). If the cuvette is repositioned so that the incident beam hits its surface at an oblique angle (Fig. 9b), the scattering pathway will be greatly reduced. This is particularly the case if a short path-length cell is used or a baffle plate is fitted to shield the bulk of the cell from the detection monochromator. This cuvette arrangement, referred to as front-surface collection, is particularly valuable for use with highly scattering samples. When using this arrangement, however, care must be taken to ensure that the reflected excitation beam is directed well away from the light-collection system.

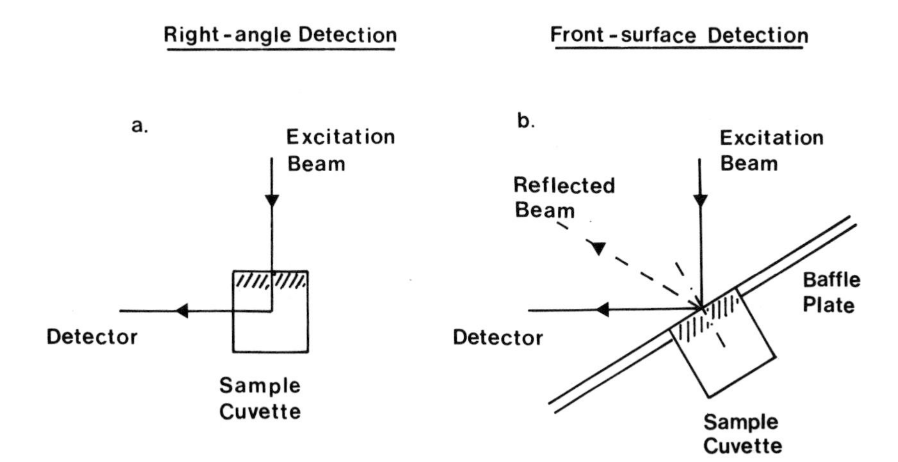

Fig. 9 Diagrams showing right-angle and front-surface detection arrangements. The hatched regions correspond to the regions in which the bulk of light absorption takes place in concentrated solutions

2.4.5 Inner-filter and Fluorescence Reabsorption Effects

The quantum yield of fluorescence is proportional to the number of photons absorbed by the emitting species. The presence of strongly absorbing species that can compete with the fluorescing species for excitation energy may lead to marked reductions in measured fluorescence yields. Inner-filter effects of this type are a particular problem in right-angle collection systems where fluorescence is collected most efficiently from the centre of the sample cuvette.

Fluorescence reabsorption occurs when emitted light is reabsorbed in its passage through the cuvette to the detection system. This can occur as a result of self-absorption or absorption by another species. Inner-filter and reabsorption effects of this type are notoriously difficult to correct for and every effort should be made to avoid them. If their occurrence is suspected, it can normally be confirmed by comparison with measurements performed on diluted samples in which such effects are reduced. In practice, artefacts of this type can best be avoided by the use of dilute samples or front-surface collection techniques.

2.5 Applications

2.5.1 Fluorescence Assay Techniques

With the increasing availability of reliable, relatively inexpensive fluorimeters and spectrofluorimeters, the use of fluorescent assay techniques has greatly increased over recent years. As an analytical method, fluorescence measurements have two major advantages over the more commonly employed absorption measurements. The first is their much greater specificity. Relatively few molecular species are naturally fluorescent; this means that any given assay mixture is unlikely to contain more than one fluorescent species. The problem of interfering species that mask the component of interest is thus much less of a problem than in assays based on absorption measurements. The second major advantage is the extremely high sensitivity that is achievable in fluorescence measurements.

Absorption measurements involve a comparison of the intensity of the light transmitted by the sample and by a suitable reference. Fluorescence measurements, in contrast, involve a direct measurement of emitted light. Absolute measurements of this type are inherently much simpler and more accurate. Whilst it is difficult to measure optical density changes below 0.01–0.001, which for most substances correspond to changes in the micromolar to millimolar range, fluorescent measurements can normally be made down to the nanomolar, and in some cases down to the picomolar, range.

The first aim in any fluorescence assay technique is to isolate, or prepare, a fluorescent derivative of the compound of interest. An enormous range of fluorescence assays is available. Some of these use the intrinsic fluorescence of the molecule of interest; others are based on the formation of fluorescent conjugates. Methods are available for the assay of amino acids, proteins, coenzymes, vitamins, carbohydrates, steroids and a large number of other molecules of biological importance. Details of many of these are provided in the references at the end of the chapter and need not be expanded upon. The second, and equally important aim, is to ensure that the measured fluorescence gives an accurate estimate of the amount of the species present in the sample. It is this latter problem that we shall concentrate on here.

Fluorescence assay techniques are normally based on the construction of a standard curve showing the fluorescence yield of known amounts of the substance of interest. This curve is then used to estimate the amount of the substance in the test samples. Inner-filter effects can, as explained in Section 2.4.5, lead to marked variations in the spatial distribution of emission in the fluorimeter cell. These, in turn, can lead to marked changes in the shape of the standard curve. Other factors that can affect the linearity of this curve are fluorescence reabsorption and concentration quenching.

In order to ensure a linear, or near-linear, relationship between fluorescence yield and the concentration of the fluorescent species, it is normally necessary to measure at very low concentrations. Fluorescence measurements made under these conditions are, however, particularly susceptible to artefacts arising from fluorescent impurities. It is thus important to use the highest quality spectroscopically pure solvents. Glass and quartz fluorimeter cuvettes and optical filters can themselves give rise to very low levels of fluorescence. Measurements should, therefore, be made on suitable blank samples to ensure that extraneous fluorescence from such sources does not interfere with the assay.

Measurements in dilute solutions can give rise to a number of problems associated with adsorption of the fluorescent species on glass surfaces, oxidation and photodecomposition (Udenfriend, 1962). The two most common sources of error in such measurements, however, are light scattering and fluorescence quenching. Even small amounts of colloidal material can lead to substantial errors. It is, therefore, extremely important to ensure that the samples used in fluorescence assays are as optically pure as possible. The presence of small amounts of fluorescence quenchers in the test samples can give rise to major problems that may be difficult to eliminate. The best method of testing for such impurities is the use of an internal standard. Addition of a known amount of the assay substance should lead to a known increase in fluorescence yield. If the anticipated fluorescence increase does not occur, the possibility of quenching should be investigated.

A final problem that should be mentioned is the possibility of artefacts arising from other fluorescent species present in the assay mixture. It is common practice in most assay techniques to measure at a fixed emission wavelength, usually the emission maximum of the material of interest. This procedure is quite acceptable as long as no other emitting species are present. It is, however, much better practice to scan the emission band of each sample. By scanning the spectrum, unanticipated problems associated with the presence of emitting species that might otherwise remain undetected can often be avoided. If such problems do arise they can often be circumvented by changes in excitation or detection wavelengths and/or slit widths. If, even after such changes, problems still remain other more sophisticated techniques such as wavelength modulation (O'Haver, 1976) and the measurement of time-resolved spectra (Beddard, 1981) are available. These methods, however, involve the use of considerably more sophisticated equipment and are beyond the scope of this chapter.

2.5.2 Fluorescent Probe Techniques

Probe techniques are based on the use of small molecules to report on the structural and/or motional characteristics of their immediate environment. Probes can be conveniently divided into two classes: intrinsic and extrinsic. Intrinsic probes are molecules that are native to the system under study. Tyrosine or tryptophan residues in proteins, or chlorophyll *a* in the photosynthetic membranes of chloroplasts are good examples. Extrinsic probes, in contrast, are foreign to the system and must be specifically added. The majority of fluorescent probes in common use are extrinsic in nature. Both types of probes have advantages and disadvantages. As intrinsic probes are native to the system, they do not perturb the environment under study: this can be a serious problem with some extrinsic probes. On the other hand, the fluorescent properties of a given intrinsic probe may not be particularly sensitive to environmental factors. Extrinsic probes, in contrast, can be chosen (or designed) to be particularly sensitive to such factors.

A wide range of fluorescent probes have been developed for different experimental purposes. Most are relatively simple dye molecules that do not interact directly with the system. In some cases, however, it is more convenient to use fluorescently labelled molecules in which the fluorescent species has been covalently linked to one of the components of the system. The major categories of probes in current use are hydrophobicity probes; viscosity or fluidity probes; membrane potential probes and energy-transfer probes. We shall examine each of these groups in turn.

2.5.2.1 Hydrophobicity Probes

These probes are characterized by the fact that their emission properties are influenced by the polarity of their environment. They show solvent-induced spectral shifts and/or solvent-induced changes in their quantum yields of fluorescence and fluorescence lifetimes. Probes of this type have been widely used to study hydrophobic clefts in proteins or to estimate the hydrophobicity of membrane binding sites (Azzi, 1975).

Solvent-induced shifts reflect differences in the extent of dipole–dipole interactions between the solvent and the probe molecule when it is in its ground and excited states. Such interactions can involve permanent dipoles or, depending on the polarizability of the solvent, induced dipoles. Molecules showing large solvent-induced shifts in emission are characterized by large differences in the size of the dipole moments associated with their ground and excited states. The degree of solute–solvent interaction and hence the extent to which such interactions stabilize the system thus differs in the two states. The origin of solvent-induced quantum yield changes is less clearly understood. The main factor involved appears to be changes in the rate of intersystem crossing associated with differences in the singlet–triplet energy separations. For a detailed discussion of these processes, see Radda (1975).

A large number of hydrophobicity probes are available but by far the most commonly used is 8-anilino-1-naphthalene sulphonate (ANS). It shows a marked blue-shift as the polarity of its environment decreases but more importantly the quantum yield of the probe increases from a value of 0.004 in water to 0.98 when it is bound to protein. This very large increase in quantum yield means that the fluorescence of unbound ANS in the aqueous phase can normally be neglected, thus greatly simplifying measurements in mixed-phase systems.

An important feature of probes of this type is that they are usually amphipathic and hence tend to partition between polar and non-polar phases. This partition, in the case of biological systems, is often influenced by the presence of charged groups at the interface between such phases. It is important to distinguish between fluorescence increases that arise as a result of an increased partition of the probe into a non-polar environment and increases that arise due to changes in the polarity of the binding site. To some extent this can be achieved by measurements of the position of the emission maximum. Increased partition (or binding) unlike polarity changes should not be accompanied by solvent-induced shifts in emission. It is, however, advisable to supplement such measurements by lifetime studies and direct binding studies wherever possible.

2.5.2.2 Viscosity or Fluidity Probes

A major group of probe techniques involves the use of probes to report on the local viscosity of their environment. Probes of this type have been widely used in membrane studies. Although hydrophobicity probes show some dependence on the local viscosity of their environment, these effects are usually quite small. Most probe techniques of this type exploit the photoselective aspects of spectroscopic transitions and use the diffusional motion of the probe molecule to monitor local viscosity. The main criteria involved in the choice of suitable probes for such studies are structural rigidity, so that the molecule moves as a single unit, and an absence of specific interactions with the surrounding medium.

All transitions between different electronic states, as explained in Section 2.2.1, involve changes in transition moments. The coupling between the oscillating electric field of the exciting light and the transition dipole of a given molecule is dependent on their mutual orientation. The more closely the two are aligned, the greater the probability of absorption. If the orientation of the molecules remains fixed with respect to the excitation beam, the fluorescence they emit will show a similar polarization to that associated with absorption. If, on the other hand, the molecules can reorientate during the lifetime of the excited state, the resulting emission will be depolarized with respect to the excitation beam. Measurements of fluorescence polarization can thus yield valuable information on probe mobility and/or the viscosity of the probe environment.

Fluorescence polarization studies of this type have been used in a number of different contexts. The earliest studies involved the use of proteins labelled with dyes such as fluorescein isothiocyanate to study the rotational motion of proteins in solution (Weber, 1953). A modification of this approach, which has excited considerable interest in recent years, is the use of triplet probe techniques to study the lateral diffusion of proteins in biomembranes. Cherry (1979) has proved an excellent account of these and related techniques. Another major field of application of photoselection techniques of this sort has been the use of molecules such as 1,6-diphenyl-1,3,5-hexatriene (DPH) and perylene to probe the fluidity of the lipid core of the biological membranes (Shinitzky and Barenholtz, 1978).

A number of different expressions have been derived relating the microviscosity (η_m) of the probe environment and fluorescence polarization. The most commonly employed in membrane studies is:

$$\frac{r_0}{r} = 1 + \frac{kT}{\eta_m V_r} \tag{6}$$

where r_0 and r are the fluorescence anisotropy of the probe in the immobilized and mobilized states respectively, k is the Boltzmann constant, T the absolute temperature and V_r is the effective molecular volume of the rotating probe. In recent years it has become increasingly clear that the concept of microviscosity, in membrane terms at least, is of dubious value in view of the highly anisotropic nature of the system and there has been an increasing tendency to present membrane fluidity data simply in the form of plots of fluorescence polarization against temperature. The use of such plots for studying the effect of gel-to-liquid crystal phase transitions in lipid bilayers is illustrated in Fig. 10.

2.5.2.3 Membrane Potential Probes

A number of fluorescent probe systems are available for the quantitative and semiquantitative measurements of changes in the membrane potential of cells, organelles and vesicles that are too small to allow the use of microelectrodes. The naphthylamine dyes, particularly ANS, have been widely used in this context (Conti, 1975). A typical trace showing the changes in fluorescence yield accompanying the energization of submitochondrial particles following the addition of succinate is shown in Fig. 11. The changes are large and are easily monitored but the method suffers from the disadvantages that they occur on a much slower time scale than the electrical changes and are difficult to quantify.

The use of naphthylamine dyes for studying membrane potential changes has now been largely superseded by the introduction of other more sensitive and more easily calibrated probes such as the cyanine, merocyanine and oxonol dyes (Waggoner, 1976). These dyes divide into two classes permeant probes (mainly cyanine and oxonol dyes with delocalized charges) and impermeant probes (merocyanine dyes with localized charges). The two groups differ both in the size and time scale of their fluorescence responses and the mechanisms underlying these changes.

The permeant probes work by a potential-dependent accumulation of the dye. They partition across the membrane in response to the size of the membrane potential change. The resulting changes in concentration of dye within the cell, or vesicle, lead to changes in its degree of aggregation and hence to its fluorescence properties. These changes, like those of the naphthylamine

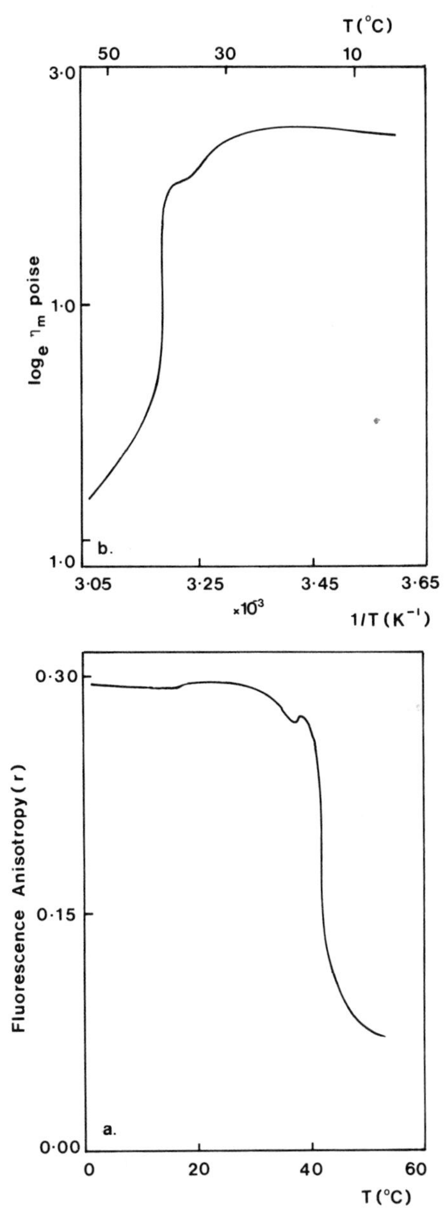

Fig. 10 Temperature profiles of fluorescence polarization anisotropy (r) and microviscosity (η_m) for multilamellar liposomes of L-palmitoyl phosphatidylcholine. The dramatic changes in the values of the two parameters occurring at about 41 °C correspond to the occurrence of a gel-to-liquid crystalline lipid phase transition. (Data redrawn from Shinitzky and Barenholtz, 1978)

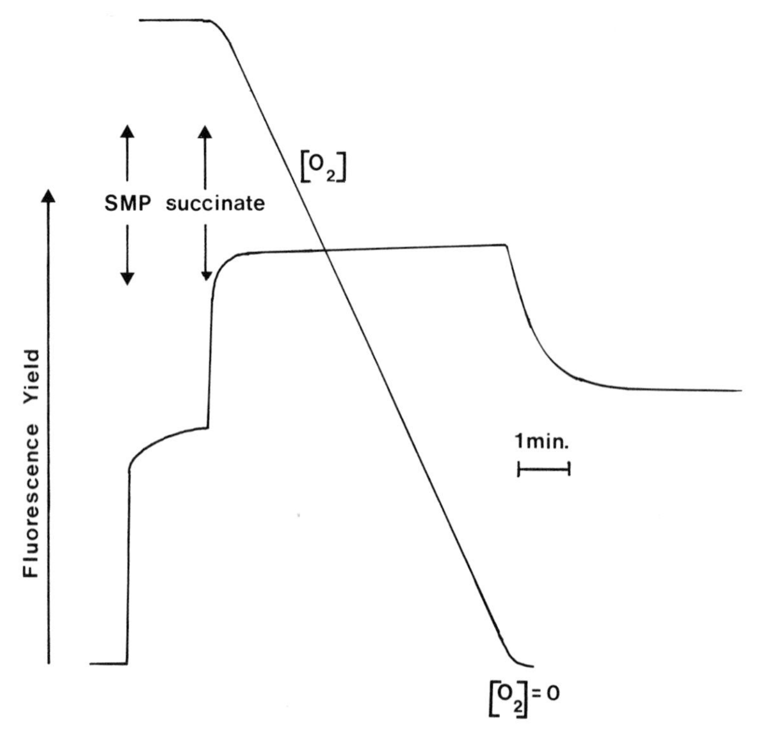

Fig. 11 Changes in ANS fluorescence yield accompanying the energization of the membranes of submitochondrial particles (SMP). The initial fluorescence increase corresponds to the binding of ANS to the membranes of the particles and the second increase to a respiration-linked energization of the particles following the addition of succinate. This latter increase is reversed when the system goes anaerobic and respiration ceases. (Based on data from Nordenbrand and Ernster, 1971)

dyes, tend to be relatively large in magnitude but slow in response. The impermeant dyes show much smaller but much more rapid responses. Typically, the fractional changes in fluorescence yield, $\Delta F/F$, are of the order 5×10^{-3} but the response time is in the submillisecond range. This, as illustrated in Fig. 12, allows the use of such probes to follow action potential changes in single cells. The mechanism underlying these changes is not fully understood but they are believed to reflect membrane-localized dye movements that lead to local changes in probe aggregation similar to those seen in the permeant probes.

2.5.2.4 Energy Transfer Probes

The use of probes of this type involves what is often referred to as the 'spectroscopic ruler' technique (Stryer, 1978). Whilst straightforward in princi-

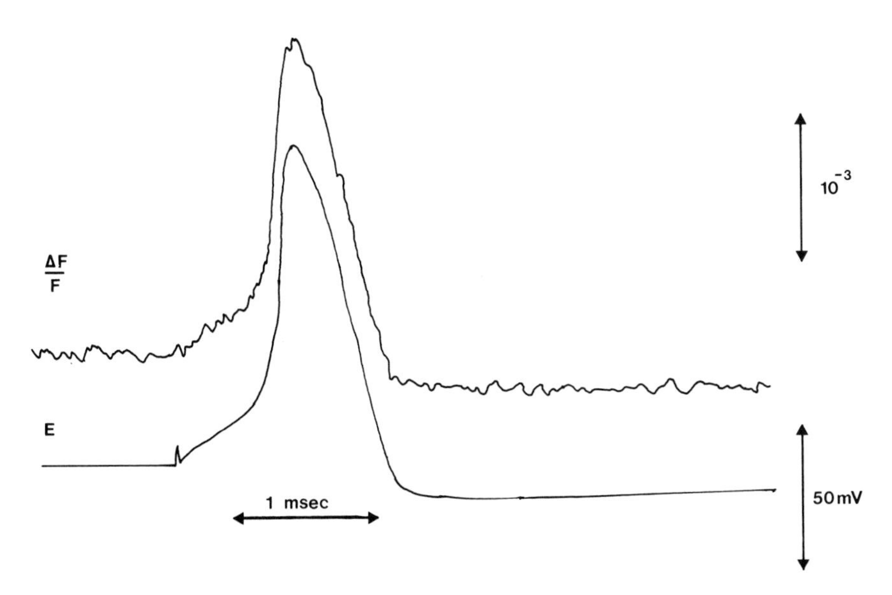

Fig. 12 Parallel fluorescence and electrical changes accompanying membrane action potentials in a squid giant axon stained with a merocyanine dye. The fluorescence increases $\Delta F/F$ represent the increase in fluorescence intensity divided by the resting fluorescence intensity. (Data redrawn from Cohen *et al.*, 1974)

ple, the method has proved rather difficult to quantify on a reliable basis. It depends on the measurement of the rate of resonance energy transfer between two chromophores; a donor and acceptor species. The rate of transfer can be followed by measurements of the fluorescence quenching of the donor species by the acceptor or by the measurement of sensitized fluorescence from the acceptor. This rate, as shown in Eq. 4, is determined by the values of the overlap integral between the corrected emission spectrum of the acceptor and absorption spectrum of the donor, the refractive index of the separating medium, the lifetime and fluorescence quantum yield of the donor in the absence of acceptor, an orientation factor and the separation between the donor and acceptor.

Two main problems have emerged. The first of these relates to the difficulty in determining the appropriate value of the orientation factor. In principle it can take any value between 0 and 2. For most purposes the value for random orientation ($K = \sqrt{2/3}$) is employed. Methods are available for bracketing this value more closely (Gennis and Cantor, 1972; Eisinger and Dale, 1974) but uncertainties regarding its precise value in any given system have proved a major obstacle in the application of this technique.

The second problem, which is particularly important in membrane studies, is that the original equations were formulated for a random, three-dimensional distribution of acceptor and donor molecules. In order to apply them to systems such as cell surfaces which are essentially two-dimensional, the equations require extensive modification. Again, some progress in this direction has been achieved (Dale *et al.*, 1981) but problems still remain.

2.5.3 Bioluminescence, Immunofluorescence and Fast Photobleaching Recovery

No account of the use of fluorescent measurements in biology would be complete without some mention of these techniques. They differ from those described above in that they do not normally involve the use of conventional fluorimetric and spectrofluorimetric equipment. They are, nevertheless, of major importance.

Bioluminescence, as we have already noted, involves the formation of excited states by chemical rather than optical excitation. A wide range of organisms ranging from bacteria, jelly fish and small crustacea to insects and fish exhibit bioluminescence. Light produced from such organisms can be measured in many ways. Micro-organisms and/or isolated bioluminescent materials are most commonly examined using a fluorescence microscope but any conventional photodetection apparatus can be used. Larger organisms require more complex apparatus. In addition to the intrinsic interest of the phenomenon itself, bioluminescent materials are used in a number of important assay systems. Of particular importance are the use of luciferase in the measurement of low levels of ATP (Thore, 1980) and aqueorin to measure Ca^{2+} levels (Blinks, 1978).

Immunofluorescence is a field of great current interest. The use of fluorescent labelled antibodies has proved to be one of the major techniques in immunology. Immunofluorescence techniques are currently used in many areas of microbiology, histology, pathology and allerology for the identification of aetiological agents in tissues and cells. They have proved extremely useful in the rapid diagnosis of many viral, bacterial and protozoan infections. A comprehensive survey of such applications has recently been provided by Kawamura (1977).

Fast photobleaching recovery is a novel technique with considerable potential use in membrane studies (Cherry, 1979). It is based on the use of fluorescent dyes to tag proteins of interest. A powerful laser is used to photobleach the labelled proteins in a small patch of the membrane of a single cell. The diffusion of other, unbleached molecules back into this area is then monitored using an attenuated laser beam. Measurements of this diffusion rate can then be used to estimate the mobility of the proteins in the membrane phase.

References

Azzi, A. (1975) The application of fluorescent probes in membrane studies. *Q. Rev. Biophys.* **2**, 237–316

Beddard, G. S. (1981) Fast time-resolved fluorescence measurements and their application to some problems of biological interest. In: *Fluorescence Probes* (eds G. S. Beddard and M. A. West), pp. 21–38. Academic Press, London and New York

Blinks, J. R. (1978) Measurement of calcium ion concentrations with photoproteins. *Ann. N.Y. Acad. Sci.* **307**, 71–85

Cherry, R. J. (1979) Rotational and lateral diffusion of membrane proteins. *Biochim. Biophys. Acta* **559**, 289–327

Cohen, L. B., Salzberg, B. M., Davila, H. V., Ross, W. N., Landowne, D., Waggoner, A. S., and Wang, C. H. (1974) Changes in axon fluorescence during activity; molecular probes of membrane potential. *J. Memb. Biol.* **19**, 1–36

Conti, F. (1975) Fluorescent probes in nerve membranes. *Ann. Rev. Biophys. Bioeng.* **4**, 287–310

Dale, R. E., Novros, J., and Roth, S. (1981) Application of Forster long-range excitation energy transfer. In: *Fluorescence Probes* (eds G. S. Beddard and M. A. West), pp. 159–81. Academic Press, London and New York

Eisinger, J., and Dale, R. E. (1974) Interpretation of intramolecular energy transfer experiments. *J. Mol. Biol.* **84**, 643–647

Gennis, R. B., and Cantor, C. R. (1972) Use of nonspecific dye labelling for singlet energy-transfer measurements in complex systems: a simple model. *Biochemistry* **11**, 2509–2517

Kawamura, A. (1977) *Fluorescent Antibody Techniques and their Applications*. Univ. of Tokyo Press, Tokyo

Nordenbrand, K., and Ernster, L. (1971) Studies of the energy-transfer system of submitochondrial particles. *Eur. J. Biochem.* **18**, 258–273

O'Haver, T. C. (1976) Modulation and derivative techniques in luminescence spectroscopy. In: *Modern Fluorescence Spectroscopy* (ed. E. L. Whery), Vol. 1. Heyden (Plenum Press), London and New York

Radda, G. K. (1975) Fluorescent probes in membrane studies. *Meth. Memb. Biol.* **4**, 97–188

Shinitzky, M., and Barenholtz, Y. (1978) Fluidity parameters of lipid regions determined by fluorescence polarisation. *Biochim. Biophys. Acta* **515**, 367–394

Stryer, L. (1978) Fluorescence energy transfer as a spectroscopic ruler. *Ann. Rev. Biochem.* **47**, 819–846

Thore, A. (1980) Luminescence in clinical analysis. *Ann. Clin. Biochem.* **19**, 359–369

Udenfriend, S. (1962) *Fluorescence Assay in Biology and Medicine*, Vol. I. Academic Press, New York and London

Waggoner, A. (1976) Optical probes of membrane potential. *J. Memb. Biol.* **27**, 317–334

Weber, G. (1953) Rotational Brownian motion and polarization of the fluorescence of solutions. *Adv. Prot. Chem.* **8**, 415–459

General Reading

Beddard, G. S., and West, M. A. (1981) *Fluorescent Probes*. Academic Press, London and New York

Bowen, E. J. (1968) *Luminescence in Chemistry*. Van Nostrand, London

Brand, L., and Witholt, B. (1967) Fluorescence measurements, *Methods in Enzymology* **11**, 776–856
Hercules, D. M. (1966) *Fluorescence and Phosphorescence Analysis.* Interscience, New York and London
Udenfriend, S. (1969) *Fluorescence Assay in Biology and Medicine,* Vol. II. Academic Press, New York and London
Wehry, E. L. (1976) *Modern Fluorescence Spectroscopy*, Vols I and II. Heyden (Plenum Press) London and New York

Biochemical Research Techniques
Edited by J. M. Wrigglesworth

3
Spin Labelling

GHEORGHE BENGA

Department of Cell Biology, Faculty of Medicine, Medical and Pharmaceutical Institute Cluj-Napoca, 6 Pasteur Street, Cluj-Napoca, Roumania

---

3.1 Introduction

A variety of labelling techniques have been used in biological investigations to probe the structure of biological systems and the mechanisms of biological reactions. For example fluorescent labels on proteins have given information about the conformations of biological molecules, while compounds labelled with radioactive or stable isotopes have provided the detailed mechanisms of many biological reactions. The concept of labelling a biological system with some external probe was termed the 'reporter' group technique by Burr and Koshland (1964). The essential feature of labelling is to introduce in a biological system a component which does not exist naturally in that system, but has properties for detection by a certain technique. The system can then be studied by a special technique which is able to 'see' the label.

Electron spin resonance (ESR) spectroscopy is a technique for detecting paramagnetism, the magnetic moment associated with an unpaired electron. The technique, sometimes called electron paramagnetic resonance (EPR), may be used for detecting transition metal ions and their complexes, free radicals and electron excited states.

It is clear that ESR spectroscopy can be used to study biological systems which naturally contain paramagnetic centres. However the natural occurrence of paramagnetism in biological systems is relatively low. Therefore it would be most desirable if any biological system or molecule could be made paramagnetic in a specific way. It would seem a natural extension of the labelling techniques to introduce in biological systems paramagnetic centres or spin labels, whose ESR spectra would provide information about the environment of the label.

Spin labelling refers to the use of stable free radicals as reporter groups or labels. Sometimes the term spin label is given only to the stable free radical covalently attached to biological molecules, while the stable free radicals dispersed in biological systems are named spin probes. However the majority of authors use the term spin label for all stable free radicals introduced in biological systems.

From the pioneering papers of Ohnishi and McConnell (1965) and Stone *et al.* (1965), who coined the term 'spin label', the literature of this field has grown tremendously. There are many reviews on the topic (Hamilton and McConnell, 1968; Griffith and Waggoner, 1969; McConnell and McFarland, 1970; Jost *et al.*, 1971; Jost and Griffith, 1972; Smith, 1972; Schreier *et al.*, 1978) as well as books (Likhtenstein, 1974; Berliner, 1976, 1979). The aim of this chapter is to provide a brief overall view of spin labelling ESR, concentrating on practical details with some examples of applications.

It would be well to start by mentioning briefly the kind of information that we might hope to obtain from an application of this technique:

(1) ESR spectra of spin labels are very sensitive to the rate at which the label is able to change its orientation. We can therefore evaluate the degree of molecular mobility in the environment of the label.
(2) ESR spectral parameters vary with the polarity of the solvent. We can therefore probe the hydrophobic or hydrophilic nature of the environment around the label.
(3) Spin labels can provide information about intramolecular distances.
(4) Biradical spin labels allow the study of certain intermolecular interactions since the interaction of the two ends of the biradical is a strong function of its conformation.

3.2 Theoretical Basis of Spin Labelling ESR

Since electrons possess both spin and charge they behave like magnets, in that they possess a magnetic moment. In a strong magnetic field unpaired electrons can exist in two states: either aligned parallel with the field in a low energy state or antiparallel in a high energy state (Fig. 1). Transitions between these energy states may be induced by application of electromagnetic radiation of the appropriate quantum of energy. If the electron is unpaired it is possible to apply electromagnetic radiation to make spin reversal (resonance) occur if the following relationship between the magnetic field and the required frequency of radiation is fulfilled:

$$h\nu = g\beta H \tag{1}$$

where h is Planck's constant; ν is the frequency of the applied radiation; g is a constant, the spectroscopic splitting factor; β is the magnetic moment of the electron, the Bohr–Procopiu magneton; and H is the magnitude of the applied magnetic field. In a magnetic field of the order of one tesla (10^4 gauss) the appropriate quantum of energy is obtained from radiation in the microwave region of the electromagnetic spectrum.

Eq. 1 indicates that the frequency of the microwave radiation absorbed is a function of the paramagnetic species (β) and the applied magnetic field. However it is usually easier in practice to irradiate the sample with a constant microwave frequency appropriate to the species being studied and vary the applied magnetic field until resonance occurs, resulting in a peak absorption of the microwave frequency (Fig. 2). Such a peak in an ESR spectrum corresponds to a paramagnetic species. The area under the peak is a measure of the concentration of that species in the sample which may be quantitatively determined if a standard containing a known concentration of unpaired electrons is available.

ESR spectra become more complex and consequently more useful, when the unpaired electron can interact with a magnetic nucleus as is often the case in spin labels. For example, ^{14}N has a nuclear spin of one unit and the magnetic moment of ^{14}N can be aligned parallel, antiparallel or perpendicular to the magnetic moment of the unpaired electron. Thus, each of the two electron energy levels is split into three. From the six ESR transitions apparently possible only those which result in a change in the polarization of the nuclear moment are observable (Fig. 3). Thus, within a first order approximation one

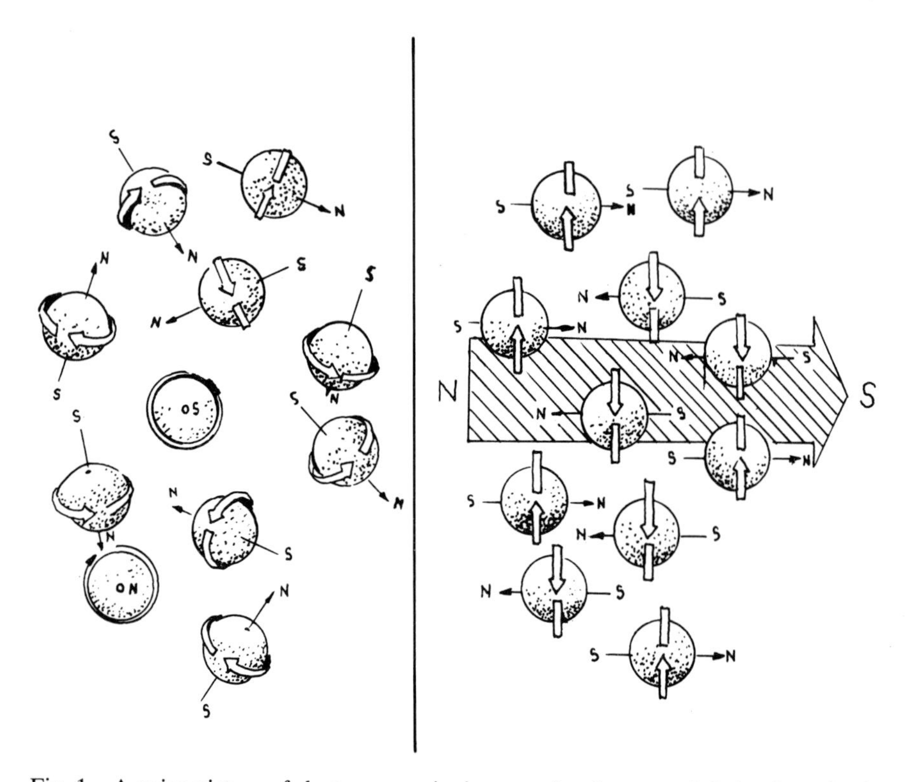

Fig. 1 A naive picture of electrons as spinning negative charges and their orientation in a strong magnetic field. Since a moving charge generates a magnetic field, the axis of each spinning electron has a magnetic moment associated with it, and the electron can be said to act like a small bar magnet with a north and south pole. (From Swartz *et al.*, 1972, with permission)

would observe three ESR absorptions, whose separations (A) are equal and represent the hyperfine splitting of the nitrogen nucleus.

The g-value is obtained from the ESR spectrum by measuring accurately the magnetic field at the centre of the spectrum and substituting the value in Eq. 1 together with the value for the microwave frequency. For a completely delocalized electron $g = 2.0023$ (Williams and Wilson, 1979).

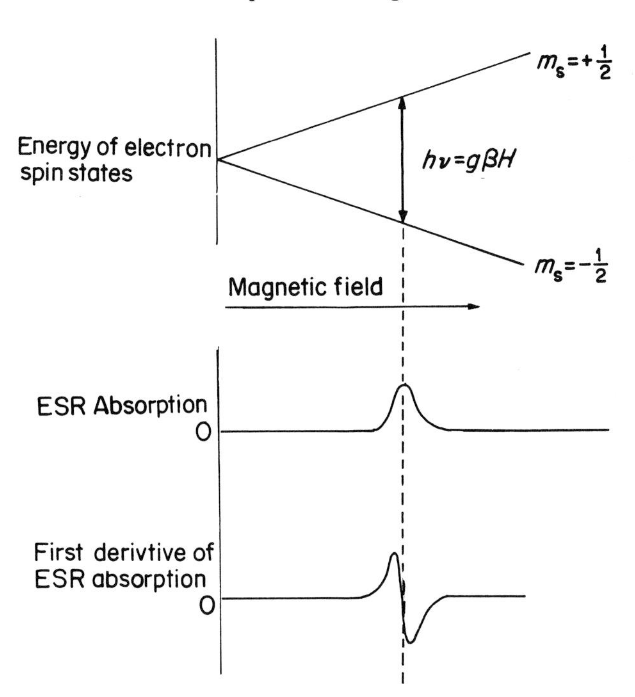

Fig. 2 The ESR experiment for a free electron. The magnitude of the magnetic field H is varied until the energy gap between electron spin states is equivalent to the energy of the microwave radiation $h\nu$. At this point absorption of microwave energy will occur. The quantum numbers $m_s = \pm \frac{1}{2}$ designate the value of the electron spin angular momentum in the direction of the applied magnetic field. (From Benga, 1979, reproduced by permission of Dacia Publishing House and the author)

3.3 Spin Labelling in Practice

3.3.1 Chemistry of Spin Labels

Although the first spin label experiment was done using the chlorpromazine radical cation intercalated within DNA (Ohnishi and McConnell, 1965) spin labels are now usually molecules containing the nitroxide moiety, which contain an unpaired electron localized on the nitrogen and oxygen atoms.

The chemistry and synthesis of nitroxides as stable free radicals has been described in several books and articles (Forrester *et al.*, 1968; Rozantsev and Sholle, 1971: Gaffney and Lin, 1976; Berliner, 1976).

There are several requirements for a nitroxide to be a useful spin label:

(1) stability under conditions used for the study of biological molecules, for example aqueous solutions, pH 2–10, high and low salt concentrations and temperatures from 20 to 70 °C;

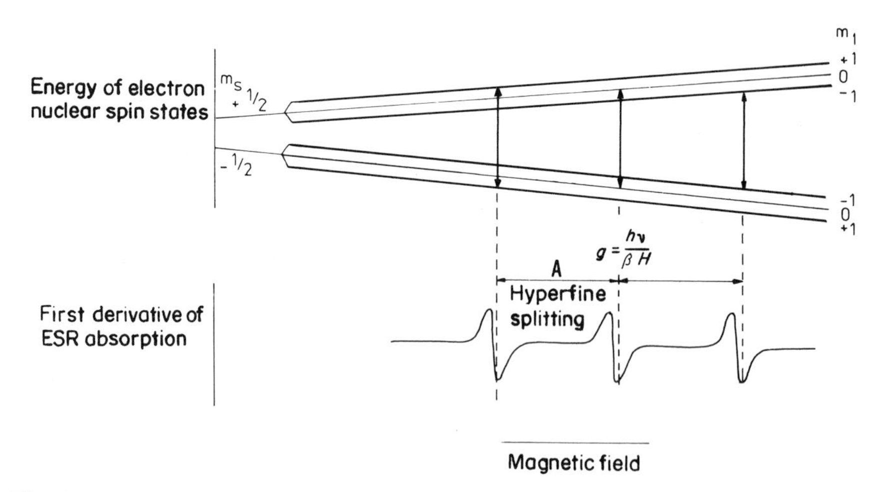

Fig. 3 The ESR experiment for a compound containing an unpaired electron influenced by a nucleus of spin 1 (^{14}N). The magnitude of the magnetic field felt by the electron will now depend on the magnetic field from the nearby nucleus ($m_I = +1, 0, -1$) as well as the applied magnetic field. For a fixed microwave frequency ν, scanning the magnetic field H will result in hyperfine splitting (A) of the signal. (From Benga, 1979, reproduced by permission of Dacia Publishing House and the author)

(2) sensitivity to its environment, preferably to polarity, acidity, spatial restriction and fluidity;
(3) the ESR spectra must be relatively simple and easy to interpret;
(4) the label should have a well-understood chemistry that permits custom synthesis and labelling of a particular site in a biological system.

Five classes of nitroxide radicals are especially useful for spin label studies (Fig. 4). The first two are based on piperidine and pyrrolidine radical systems and were originally developed by Rozantsev and Sholle (1970). Later, other nitroxide radicals were introduced where substitutions can occur quite close to the N–O group. To this category belong doxyl nitroxides, proxyl nitroxides and azethoxyl nitroxides. Methyl groups afford stability to the spin labels. Although nitroxides are extraordinarily stable free radicals, they may be destroyed, with loss of paramagnetism, by components of some biological systems and under some of the experimental conditions employed in synthetic steps.

For a biologist or biochemist it is more useful to know what can be used for R in the structures shown in Fig. 4. One may distinguish several types of spin labels and spin probes. Examples are shown in Fig. 5.

(1) Tempo and its derivatives (I–IV).
(2) Spin labelled protein modification reagents, like maleimide label (V) which reacts with sulphhydryl and amino groups, or iodoacetamide spin

label (VI) which reacts with sulphhydryls more slowly than the maleimide label and in some cases this may influence the specificity achieved in labelling sulphhydryl versus amino groups. A new spin label for SH groups in proteins has been recently synthesized (Sholle *et al.*, 1980).

(3) Spin labelled lipids: derivatives of steroids (VII, VIII) fatty acids (IX) and phospholipids (X, XI).

(4) Spin labelled cofactors and prosthetic groups: derivatives of NAD (XIII) corrinoides and vitamin B_{12} and of the haems of haemoglobin, cytochrome *c* (XIV).

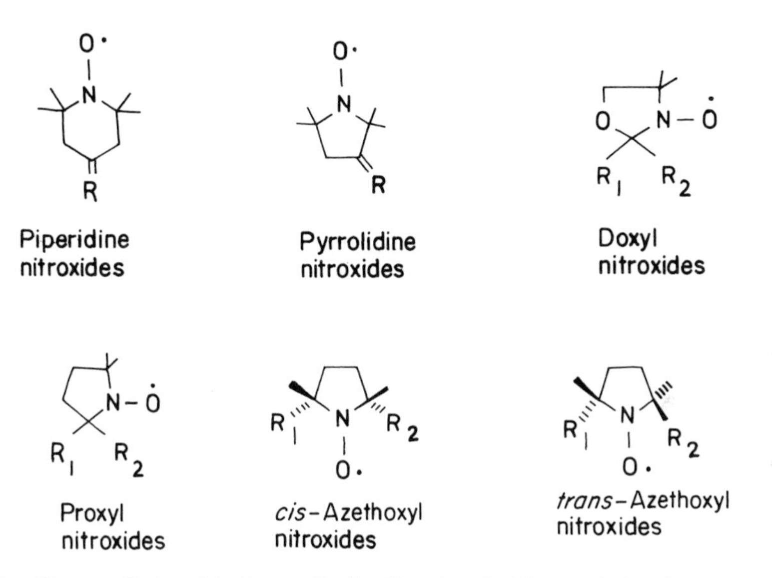

Fig. 4 Classes of nitroxide free radicals. (Reprinted with permission from Berliner, 1981, copyright The Chemical Rubber Co., CRC Press Inc.)

(5) Spin labelled nucleotides: derivatives of AMP (XII) or ATP.

(6) Spin labelled sugars: derivatives of galactose, *N*-acetylglucosamine (ester) (XV) and thiogalactose.

(7) Spin label probes of binding site structure: spin labelled nitrophenyl ester substrates for α-chymotrypsin, sulphonamide labels for carbonic anhydrase and nitrophenyl spin label antigens provide good examples of reagents with a variable chain length between the 'reporter' group and the group bound by the protein.

(8) Spin labels in immunochemistry: spin labelled nitrophenyl haptens. Many of these compounds can be obtained from commercial sources (see Appendix to this chapter).

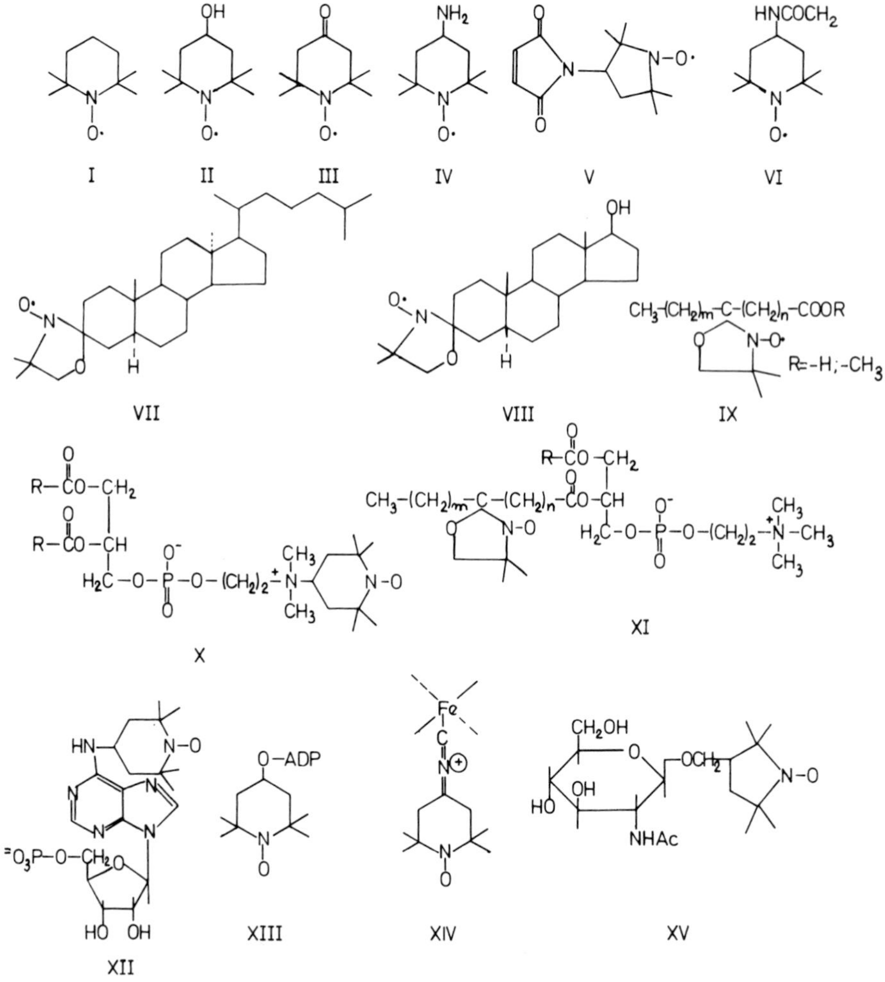

Fig. 5 Some examples of spin labels. I. Tempo (2,2,6,6-tetramethyl-piperidin-*N*-oxyl). II. Tempol. III. Tempone. IV. Tempamine. V. Maleimide spin label. VI. Iodoacetamide spin label. VII. Cholestan spin label (3-doxyl cholestan). VIII. Androstan spin label (3-doxyl androstanol). IX. Fatty acid spin labels. X, XI. Phospholipid spin labels. XII. Spin labelled AMP. XIII. Spin labelled NAD. XIV. Spin labelled haem. XV. Spin labelled *N*-acetylglucosamine

3.3.2 Practical Aspects of Labelling of Biological Samples

Most spin labels are solid at room temperature, an exception being the methyl esters of spin labelled fatty acids which are viscous liquids.

For labelling a biological sample one should take into account the nature of both sample and spin label. In some cases a small amount of solid spin label can be added to a sample dissolved or suspended in aqueous solution. For instance it is recommended for labelling with maleimide spin label to add 1.2 mg label to 0.34 g bovine serum albumin dissolved in 5 ml of 0.1 M phosphate buffer (pH 6.8). After stirring for several hours at 0 °C the solution is dialysed against buffer to remove the non-bound or free label (Griffith and McConnell, 1966). However, it may be difficult to establish when all the free label in solution has been removed (Benga and Strach, 1975). It is also difficult to estimate quantitatively the amount of label which is bound to the biological sample (Jones and Woodbury, 1978).

The commonest way of labelling is to prepare a stock solution of spin label in an organic solvent (ethanol, methanol, chloroform, etc.). The concentration of this solution is usually around 10^{-2}M. A small amount of this solution (5–10 μl) is placed on the bottom of a vial, evaporated to dryness under nitrogen and then placed in vacuum for 2–12 h to remove traces of solvent. The biological sample, dissolved or suspended in aqueous solution, is then added and mixed with the label mechanically.

The ratio between the label and sample is important and is a function of the nature of both label and sample. For example in case of a protein containing sample which is labelled with a spin labelled protein modification reagent a one to one molar ratio between the label and protein is advised. The length of incubation, the temperature and pH should correspond to those optimal for the particular protein and modification reagent. Generally an incubation of 1 h at 37 °C or overnight at 4 °C results in good labelling.

We have found it convenient to incubate for 1 h at room temperature with magnetic stirring and then to incubate the sample at 0–4 °C overnight. Excess spin label is then removed by dialysis (in case of protein solutions) or by centrifugation steps (in case of biological membrane samples). Alternatively, chromatography on a column of Sephadex G25 may be used.

For model phospholipid membranes (liposomes) there are two types of spin labels that are routinely used. Spin labels such as Tempo can be used as a 5×10^{-2}M aqueous solution, 5–10 μl of which is added to 50–100 μl sample of liposomes containing about 20 mg lipid/ml. The sample is thoroughly mixed and then closed in the sample tube to be placed in the ESR cavity. Alternatively the preparation of liposomes can be achieved by dispersing the lipids in a 5×10^{-3}M aqueous solution of Tempo.

For labelling liposomes with lipid spin labels the label and lipids are mixed together in organic solvents (chloroform, ethanol, etc.) in a label/lipid weight ratio around 1/100. For example 5–10 μl of a fatty acid or phospholipid spin label is mixed with 25–50 μl of a 20 mg/ml solution of lipid in chloroform–methanol. The solvent is then removed first under nitrogen and then in vacuum and the sample dispersed in 50–100 μl water. The dispersion should be performed at a temperature above that of the solid-to-liquid crystalline transition of phospholipids (for a description of this transition see Chapman and Wallach, 1973).

A different way of labelling biological samples is required when working with intact tissues, for example nervous tissue. The dissected nerve is soaked in an isotonic solution containing 10^{-4} M spin label and then washed in isotonic solution to remove unbound label (Giotta and Wang, 1973; Viret *et al.*, 1979). Alternatively the nerve samples may be labelled by exchange of the label with bovine serum albumin (BSA). A solution of BSA is first labelled by agitation with a spin label deposited as a thin film at the bottom of a vial. The nerve is soaked in this labelled BSA for exchange to occur and then washed to remove the BSA. In order to avoid a lowering of the (nitroxyl) spectrum intensity the labelling may be performed in the presence of 1 mM potassium ferricyanide which prevents reduction of the nitroxyl radical. Chloro-2-ethanol can replace BSA in the procedure described above (Laporte *et al.*, 1979). The exchange from BSA can also be applied to labelling of other membranes, such as those of *Mycoplasma* (Rottem *et al.*, 1970) or erythrocytes (Landsberger *et al.*, 1971).

Finally the possibility of introducing the spin label in biological systems by biosynthesis should be mentioned. This has been accomplished with spin labelled fatty acids incorporated into the membrane lipids of mitochondria (Keith *et al.*, 1968; Tourtellotte *et al.*, 1970) and microsomes (Stanacev *et al.*, 1972) and also in nucleic acids (Bobst *et al.*, 1979).

After labelling, the sample (around 50 μl) is usually drawn to the centre of a thin-walled capillary tube made of ordinary flint (soft) glass or pyrex (melting point tubes can be used) which is sealed at both ends with a flame. The sample is then displaced to one end (by shaking) and the capillary placed into a standard 2–4 mm i.d. quartz ESR sample tube. After the experiment the capillary can be discarded and another sample prepared and similarly placed in the quartz tube. Glass tubes can be used instead of quartz. However, special care must be taken to check for intrinsic ESR signals.

Aqueous samples can be studied in the same way but the dielectric loss of water tends to lower the spectrometer sensitivity by lowering the quality of the cavity. Alternatively special flat aqueous sample cells of quartz (approximate dimensions 6 cm $\times$ 1 cm $\times$ 0.3 mm i.d.) which can have stoppered ports at the top and the bottom may be used. Organic samples, powders, dry tissues and other samples that do not contain water may be drawn into large diameter flint glass or pyrex capillaries, or disposable pipettes sealed at the small end.

For particular purposes many special sample holders have been constructed. These range from a simple glass cover slip to support phospholipid multilayers, to the complex holders designed to permit nerve excitation and ESR signal recordings (Calvin *et al.*, 1969).

3.3.3 Instrumental Aspects

A schematic diagram of an ESR spectrometer is shown in Fig. 6. The basic parts are the klystron tube to generate the microwaves, the 'magic tee' junction, a cavity, a detector–amplifier combination and a recording device. The klystron in conjunction with the isolator and the adjustable attenuator provides the operator with a variable source of microwave energy. The 'magic tee' is the microwave equivalent of the physicist's Wheatstone bridge. The impedances of arms 2 and 4 are nearly matched and very little power reaches the detector on arm 3. The klystron frequency is usually fixed at $\nu = 9 \times 10^9$ Hz (3×10^{-2} m) and the magnetic field is scanned.

The sample is situated in the microwave cavity, a region of high magnetic field, H. When the resonance condition (Eq. 1) is approached, the sample absorbs microwave energy, producing an imbalance in the bridge. This imbalance is detected as increased power in arm 3 and is amplified and displayed on a recorder. The modulation coils, which are fastened to the sides of the

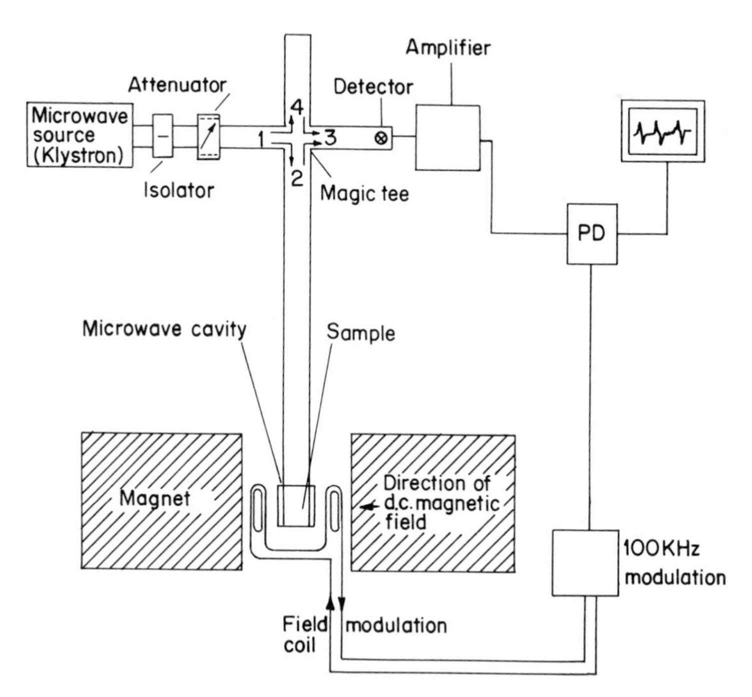

Fig. 6 A diagram of an ESR spectrometer showing the essential components. (After Jost and Griffith, 1972, reproduced by permission of Plenum Publishing Corp.)

microwave cavity, out of sight of the operator, are needed for conversion of the d.c. signal to an a.c. voltage. This can be easily amplified using the powerful methods of phase-sensitive detection. The use of modulation causes the spectra to appear as a first derivative of the microwave absorption. The modulation frequency used is 100 kHz. Most experiments are conducted around 330 millitesla (mT) in conjunction with an auxiliary sweep of 10–100 mT (1 mT≡ 10 gauss). Some commercial sources of spin labelling equipment and supplies are given in the Appendix.

Although the mode of operation of a particular ESR spectrometer is described in its instruction manual there are a number of practical details to be considered in order to record properly the ESR spectra of spin labelled biological samples and avoid instrumental artefacts.

After the sample has been introduced into the cavity and the tuning between the sample and cavity checked, the signal has to be located. Working with nitroxides a value of the magnetic field set around 3300 G and a scan range of 100 G is routinely used.

The next step is to check the set of the receiver gain, modulation amplitude and microwave power controls. If any or all three of them are set to zero, no signal will be observed on the recorder. The optimal range of the gain, modulation amplitude and microwave power depends on the sample and to a lesser extent on the ESR spectrometer. As each variable is increased, the signal amplitude increases, passes through an optimal range and then either decreases or becomes distorted. The receiver gain (also called amplifier gain) may be increased until the point is reached where the amplifier becomes unstable and the signal-to-noise ratio decreases. The modulation amplitude has a dramatic effect on the spectra. As the modulation amplitude increases, the ESR lines first increase in height, then broaden and finally become greatly distorted. A useful rule is to set the modulation amplitude equal to or less than the ESR line width (in gauss). However, if careful line shape measurements are being made, the modulation amplitude (or other instrument setting) should always be decreased and the effect on the line shape should be observed. Only if the change in line shape or relative peak heights is insignificant can the setting be considered correct.

The effects of microwave power on nitroxide ESR spectra are also important. In any system, after absorption of microwave energy there are relaxation processes that allow the spins to return to the ground state. If the power is too large, the relaxation processes are unable to return the spin system to equilibrium, and what is known as saturation takes place. Microwave power values of 1–5 mW are usually acceptable for spin labelling studies at room temperature. Once again the setting should be checked by decreasing the power and looking for changes in the ESR spectrum.

Improper adjustment of the scan time and the filter time constant produce similar line shape distortions. The scan time (or sweep time) is the time required to vary the d.c. magnetic field slowly over a specified interval (the scan range). A scan time too short will distort the ESR spectrum. ESR spectrometers have filter networks to increase the signal-to-noise ratio and the filter time constant must be much shorter than the time required to sweep through the ESR line. Generally the time constant should be one order of magnitude smaller than the scan time. However, to avoid distortion of spectra it is recommended to check the time constant, scan time and scan range experimentally.

A similar line shape distortion as that produced by improper adjustment of scan time and time constant is caused by d.c. magnetic field inhomogeneity. This can be usually corrected by adjusting the position (tuning) of the microwave cavity.

Other factors producing strong effects on the ESR spectra of spin labels are oxygen–nitroxide and nitroxide–nitroxide interactions. The ground state of molecular oxygen is a triplet and oxygen is therefore paramagnetic. It can therefore interact with the spin label through exchange and dipolar mechanisms. The result is the oxygen broadening of spectra. This is more marked in methanol, chloroform and other hydrocarbon solutions than in aqueous solutions. The oxygen may be removed by the freeze–thaw method or by bubbling a good grade nitrogen or argon gas through the samples.

Nitroxide–nitroxide also occur via dipolar and exchange mechanisms. These may occur even in the case of a solution of spin label in an organic solvent when the concentration is increased. Generally the sharp three-line spectrum remains unchanged until the concentration exceeds 10^{-3} M. At higher concentrations the three lines gradually broaden and move together until an exchange-narrowed, single-line spectrum occurs.

The reader is strongly advised to see samples of the effects of all these instrumental variables on ESR spectra in the reviews of Jost and Griffith (1972) and Bolton *et al.* (1972).

3.3.4 Interpretation and Analysis of Spectra

The aim of this chapter is not so much to give rigorous and mathematical formulations of the phenomena but rather to describe what useful information can be provided by an ESR spectrum. Those interested in a complete description of the phenomena can find this in the two volumes of the book edited by Berliner (1976, 1979).

Fig. 3 showed the origin of an ESR spectrum of a nitroxyl spin label and the variables g and T were indicated. Both g and T depend on the orientation of the paramagnetic molecule in a magnetic field (H) and the direction of H is specified in terms of a conventional axis system for nitroxides (Fig. 7).

ESR spectra of spin labels depend very much on the motion of the nitroxide group, on its orientation with respect to the applied magnetic field, on the

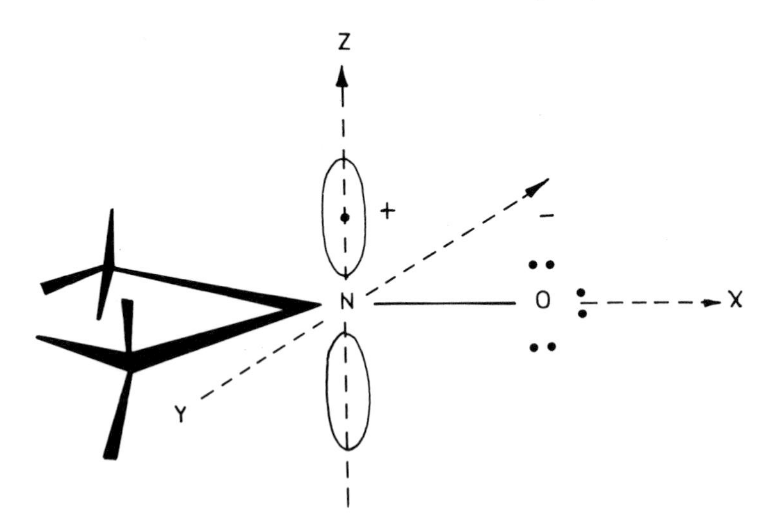

Fig. 7 Molecular axes for nitroxides. The unpaired electron is considered to occupy a molecular orbital composed of the P_π orbitals of nitrogen and oxygen. A substantial fraction of the unpaired electron is localized on the nitrogen atom and therefore the N–O three-electron π-bond has a polar character. z axis lies along the nitrogen π-orbital and x along the N–O bond. (After Smith, 1972, with permission of John Wiley & Sons Ltd)

polarity of microenvironment of the label and on the presence in the magnetic environment of nearby nitroxides or other paramagnetic centres (free radicals or paramagnetic metal ions).

3.3.4.1 ESR Spectra of Nitroxides in the Absence of Molecular Motion

There are two situations of interest:

(1) where the nitroxides are oriented in a rigid manner, for example in a crystal, and
(2) where a small concentration of nitroxide is present in a rigid glass, polycrystalline sample or powder.

The host crystal orients the nitroxide molecules so that their x, y, and z molecular axes (Fig. 7) are aligned in well defined directions relative to the crystalline axes. The crystal can be rotated until the magnetic field (H) is parallel to the x, y or z molecular axis. As shown in Fig. 8 a small splitting (5–6 G) is observed with the field parallel to x or y while a very large splitting (30–34 G) occurs along z. Because of the geometry of the molecule each direction is also associated with unique g-factors.

Advantage has been taken of anisotropy relations in determining both

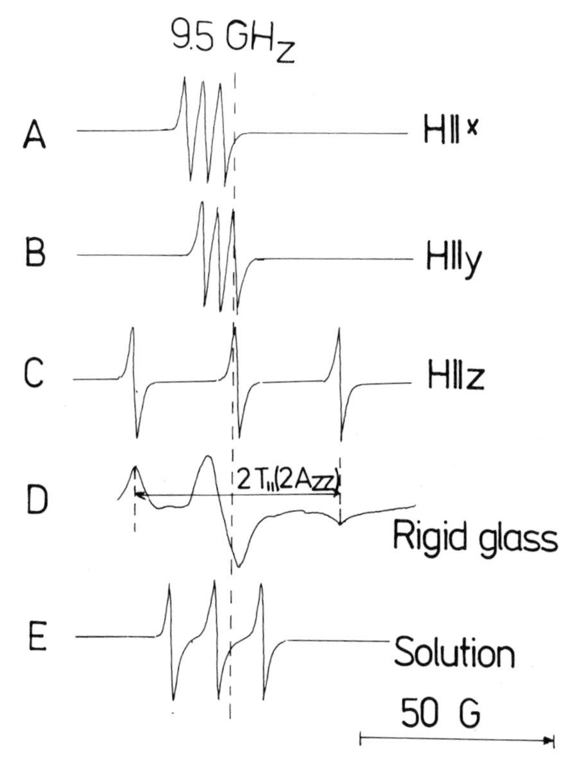

Fig. 8 Calculated spectra at 9.5 GHz with the field along each axis of a rigidly oriented nitroxide (A, B, C), a collection of randomly oriented nitroxide in the absence of motion (D), and the same collection undergoing rapid isotropic motion (E). (After Jost *et al.*, 1971, reproduced by permission of Academic Press Inc.)

precise spin label orientation and symmetry relationships in single crystals of spin labelled proteins. For example studies of α-chymotrypsin labelled with the $S(-)$ enantiomer of the spin labelled substrate has been studied by Bauer and Berliner (1979). It was possible to place exactly the nitroxide molecule at the α-chymotrypsin active site with the orientation found from the ESR experiments, so that the spin labelling technique accomplished effectively a crystallographic difference mapping of an enzyme–substrate complex (Berliner, 1981). The structural information derived from such work sheds more light on mechanistic hypotheses regarding substrate orientation and the efficiency of its subsequent catalysed hydrolysis.

If the spectra A, B, and C of Fig. 8 represent full orientation, the other exteme is the completely random orientation of nitroxides as illustrated in the lower spectra (D and E). The rigid glass or powder spectrum (D) results from a simple superposition of spectra from randomly oriented spin labels. The same sample undergoing rapid isotropic motion (E) is dealt with below.

3.3.4.2 The Effects of Molecular Motion on ESR Spectra

Isotropic motion. An isotropic motion is completely random about all of the principal axes of the nitroxide. The rapid isotropically tumbling nitroxide gives rise to a spectrum of three sharp lines (Fig. 8 spectrum E and Fig. 9 spectrum A). This is found with all spin labels in non-viscous solvents (water, organic solvents) at room temperature if the label is soluble in a particular solvent and if the concentration is not too high (so that exchange-broadening does not occur).

If the rate of molecular tumbling decreases, differential broadening of the three-line nitroxide spectrum occurs. The simplest case, that of symmetrical rotation at gradually decreasing rates, can be demonstrated by recording spectra of a small spin label like Tempol or Tempo in aqueous solutions of glycerol at various temperatures (Fig. 9). Each spectrum can be characterized

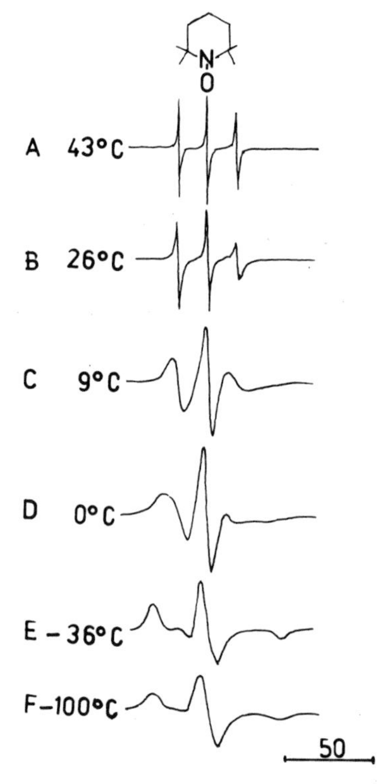

Fig. 9 The effect of viscosity on the ESR spectra of Tempo (5×10^{-4} M) in glycerol. (After Jost *et al.*, 1971, reproduced by permission of Academic Press Inc.)

by a correlation time for rotational reorientation; this rotational correlation time can be thought of as being the time required for a molecule to forget its previous orientation or the time required for a nitroxide undergoing Brownian motion to rotate through a significant arc, e.g. 40°. For the spectra in Fig. 9 the correlation times were calculated from the Stokes–Einstein equation:

$$\tau = \frac{4\pi\eta r^3}{3kT} \tag{2}$$

where η is the viscosity, r the effective radius of the molecule (which is assumed to have spherical symmetry), k the Boltzmann constant and T the absolute temperature.

This is only an estimate, since the Stokes–Einstein equation describes macroscopic viscosity, and it is in fact microviscosity that determines the correlation time for a given radical. However the spectra in Fig. 9 are illustrative for various types of ESR spectra.

The spectrum A in Fig. 9 corresponds to a very mobile or freely tumbling spin label. By decreasing temperature down to 26 °C the line positions change very little, remaining at their isotropic splittings but the relative widths change indicating differential relaxation. The spectrum B is characteristic for a nitroxide with 'slightly hindered' rotation or a weakly immobilized spin label. At lower temperatures the high and low field lines move gradually to higher and lower fields respectively. This is because the rotational reorientation rate is no longer sufficiently fast to average out the anisotropies. Spectra C and D correspond to 'moderately immobilized' nitroxides, while spectrum E is often referred to as that of a 'strongly immobilized' spin label. The limit of such immobilization results in a polycrystalline (powder) spectrum which corresponds to nitroxides possessing all possible orientations with respect to the magnetic field but being rigidly fixed. Spectra at −100 °C (spectrum F in Fig. 9) are typical 'rigid glass spectra' or 'powder spectra'. It is thus obvious that nitroxide ESR spectra are very sensitive to the rate of molecular rotation, covering a range of correlation times from 10^{-10} sec (spectrum of three narrow lines of almost equal width) to 10^{-7} sec (a strongly immobilized spin label spectrum). This range of correlation times includes those of most biological molecules, both large and small, and this explains the very general applicability of the technique to conformational problems in molecular biology. By saturation transfer ESR spectroscopy (see Section 3.5) motions with characteristic times between 10^{-7} sec up to 10^{-3} sec can be measured.

It is important in practice to know the parameters to use for estimating the mobility of spin labels. In case of fast isotropic motion (i.e. with spectra of the type A–C in Fig. 9) the rotational correlation time may be used to quantitate the rate of motion. For the time range 10^{-12} to 10^{-10} sec the theoretical treatment of Kivelson (1960) and Goldman *et al.* (1972) is applicable. Various formulae have been suggested for estimating correlation times shorter than approximately 5×10^{-9} sec (for a discussion see Schreier *et al.*, 1978; Berliner, 1981). As an experimental parameter of spin label mobility in case of fast isotropic motion the ratio of heights of low field and central line or high field and central line may be used.

When the motion of the label is restricted (for spectra of the type D, E, and F in Fig. 9) it is difficult to estimate correlation times, particularly since the motion of the spin label is anisotropic. A useful experimental parameter in these cases is the separation between the two extreme components of the ESR spectrum which can be used as a qualitative estimate of the degree of immobilization of the spin label (the so called $2T_{\parallel}$ shown in Fig. 8). A greater value of the $2T_{\parallel}$ reflects more restricted rotational motion.

This time range $\tau_c = 10^{-8}$ to 10^{-10} sec has been treated in detail by Freed (see Berliner, 1976) who has developed several computer simulation models for both isotropic and anisotropic tumbling.

There are situations in practice when complex spectra containing at least two discernible subspectra are observed. For example in Fig. 10 we have the ESR spectra of BSA labelled bovine serum albumin with maleimide spin label. Subspectra corresponding to highly constrained (strongly immobilized) and partly constrained (weakly immobilized) spin labels are evident. There are regions in the composite ESR spectra in which only one of the two types of spectrum makes a contribution to the amplitude. The ratio of these amplitudes (b/a or c/a in Fig. 10) serves as a useful qualitative monitor of conformational transitions (Benga and Strach, 1975).

Anisotropic motion. This refers to any molecular reorientation that does not occur with equal probability in all directions. In such cases asymmetric line broadening and spectral changes occur due to the contributions of the anisotropic electron–nuclear hyperfine interaction between the free electron and the ^{14}N nucleus, as well as the anisotropy in the *g*-factor.

Anisotropic rotation about one molecular axis (R) combined with motion in other direction can lead to complex line shape analyses beyond the scope of this chapter.

3.3.4.3 Solvent and Polarity Effects

Both parameters A and g (Fig. 3) are solvent dependent. The value of A decreases and g increases as the solvent polarity is decreased. The value of A is typically 1 or 2 G smaller in hydrocarbon solvents than in water and the correspondent increase in g is of the order of 0.0005. For example di-*t*-butyl nitroxide in water exhibits $A_w = 16.7$ G and $g_w = 2.0056$, whereas the same nitroxide in hexane yields $A_h = 14.8$ G and $g_h = 2.0061$ (Kawamura *et al.*, 1967).

Recalling that g determines the centre of the spectrum and A is the distance between adjacent lines, the effect of increasing g is to shift the entire spectrum to lower fields while the small decrease in A contracts the ESR spectrum slightly. Thus, when a spin label is present in both an aqueous and hydrocarbon phase, the maximum separation occurs between the two high field lines.

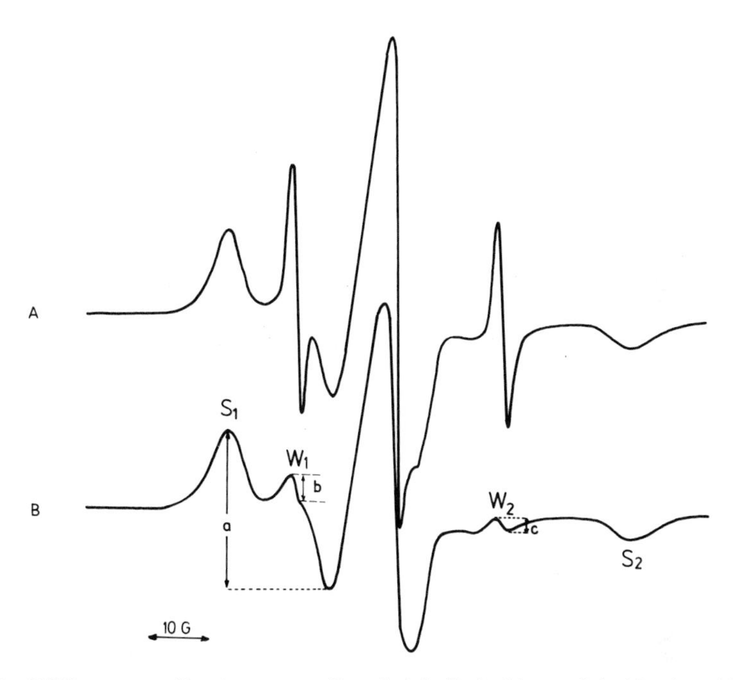

Fig. 10 ESR spectra of bovine serum albumin labelled with a maleimide nitroxide (V in Fig. 5). 1.2 mg of solid label was added to 0.34 g albumin dissolved in 5 ml of 0.1 M phosphate buffer (pH 6.8). After stirring for 3 h at 0 °C the solution was dialysed against 0.1 M phosphate buffer (pH 6.8) at 5 °C for 24 h (spectrum A) and 3 days (spectrum B). Spectra were recorded with a Varian E3 spectrometer with modulation amplitude 1.6 G, with microwave power 2 mW, and with gain 8×10^4. The strongly immobilized (S_1 and S_2) and weakly immobilized (W_1 and W_2) components are indicated. The ratio of the amplitude b/a or c/a can serve as a useful qualitative monitor of conformational transitions in the molecule

An example of such effects is provided by the spin label Tempo, a small molecule nitroxide useful in membrane studies because of its high solubility in both aqueous and hydrocarbon environments. Fig. 11 gives a spectrum which arises from Tempo partitioned between the hydrophobic region of rat microsomal membranes and the aqueous environment. To a first approximation P is proportional to the amount dissolved in the aqueous region. The fraction of Tempo dissolved in the fluid lipid phase of the sample at any one temperature is then equal to $H/H+P$ (Shimshick and McConnell, 1973).

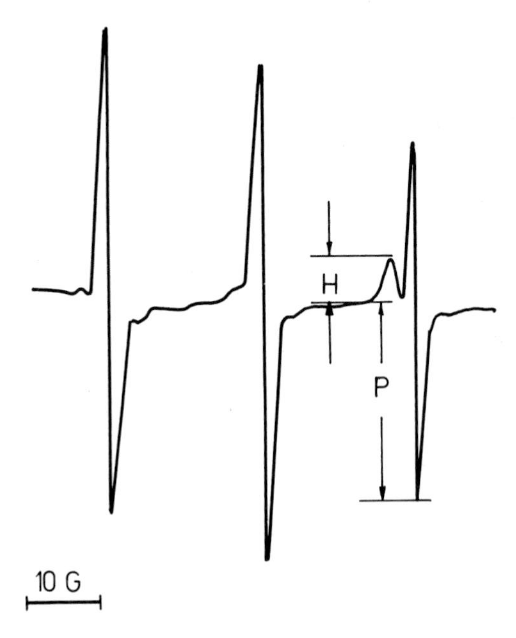

Fig. 11 A typical ESR spectrum of Tempo in a suspension of human liver microsomes (20 mg protein/ml). The spectrum was recorded with a Jeol (model 1963) ESR spectrometer at 43 °C. The Tempo partitions between the hydrophobic region of rat microsomal membranes and the aqueous environment. To a first approximation P is proportional to the aqueous concentration, and $H/H+P$ proportional to the fraction of Tempo dissolved in the fluid lipid phase

3.3.5 The Use of Computers in ESR

Several small computers appropriate for interfacing with spectrometers are commercially available. The computer controls the ESR spectrometer through commands from the teletype, and digitalizes the data as the spectrum is collected. Data treatments can be performed either as the points are collected or immediately after. The data can also be stored and later re-entered into the computer for processing.

Some of the applications of the use of small computers in ESR are listed below.

(1) *Time averaging* improves the signal-to-noise ratio in a very noisy spectrum by repeated scanning. The change in signal-to-noise ratio is proportional to the square root of the number of scans. It is always preferable to increase the signal-to-noise ratio by introducing more label or concentrating the sample but, if this is not possible, time averaging can be very useful in improving the quality of the spectrum and revealing otherwise inaccessible details.

(2) *Spectral titration* is very useful for analysing composite ESR spectra, when the spin label is partitioned into two environments, for example spectra arising from equilibrium concentrations of free and bound spin label. In spectral titration by summation, difference spectra are obtained by subtraction of one component from the composite experimental spectrum. The end-point is judged by the disappearance of that component, by the overall line shape, and by the appearance of phase reversal of the component. The estimation of the relative proportion of the mobile and immobile components in a composite spectrum may be carried out by spectral titration against either of the component spectra.

(3) *Summation* of the mobile and immobile components in different proportions to match the experimental composite spectra is a second method of spectral analysis. Examples of applications of the (2) and (3) type may be found in Jost *et al.* (1973) and Benga *et al.* (1981).

(4) *Integration* of spectra may be used for concentration determinations. Since the ESR spectrum is recorded as the first derivative of the absorption spectrum, integrating once yields the absorption spectrum. A second integration gives concentration.

(5) *Simulation* of spectra by computer has been recommended for analysis of pseudo-isotropic spectra for which it is not possible to extract correlation times. The approach has been applied by Cannon *et al.* (1975).

3.4 Examples of Biological Applications

The following sections present selected examples of biological applications of ESR in order to provide some idea of the possibilities of the method and perhaps to suggest ways of using spin labelling in particular fields of research.

3.4.1 Nucleic Acids

Early investigations (reviewed by Smith *et al.*, 1976) have shown that alkylating spin labels, such as analogues of *N*-ethylmaleimide, bromoacetic acid and bromoacetamide, react with the bases in both DNA and RNA. Studies of the order–disorder transitions for polyadenylic acid and polyguanylic acid were made by variations in pH; the resultant pK values agreed with those obtained by optical spectroscopic methods. More recent melting studies with polyadenylic acid and polyuridylic acid have shown that conformational transitions induced by temperature may be analysed by ESR provided a low nitroxide to nucleotide ratio is used, for example 0.002 (Bobst *et al.*, 1979).

Hoffman *et al.* (1969) reported studies on spin labelled tRNA. Plots of log τ versus the reciprocal of absolute temperature showed discontinuous behaviour. The temperature at which this discontinuity occurred was somewhat lower than the melting temperature obtained from the more conventional thermal-optical profiles and also depended on the nature of the solvent and the ionic strength. The findings were considered to indicate that two molecular states of tRNA were possible, each of which differed in the degree of constraint of the spin label.

Spin labels have been used to investigate histone–DNA interactions. Lawrence *et al.* (1980) have performed the labelling of histone H1 in two ways:

(1) covalent labelling of the ε-amino groups of basic amino acids (lysine and arginine) which are distributed on both ends of the molecule in the primary sequence;
(2) labelling of hydroxylated residues (serine, threonine and tyrosine) which are found in the central globular part of the molecule (in the middle of the primary structure).

When the labelled histone was free in solution it exhibited an ESR spectrum characteristic of a label moving in a viscous medium. When the histone labelled at ε-amino groups interacted with DNA the label became fully immobilized. In contrast the histone labelled on the OH groups failed to reveal a strong immobilization of the label upon interaction with DNA. These results were considered to suggest that the ε-amino groups involved in the interaction may represent anchorage points in the binding process of the histone H1 to DNA, while the rest of the molecule would interact rather weakly with DNA.

Spin labelling studies of the tyrosyl residues in the nucleosome core have been performed in order to investigate the accessibility and conformational state of core amino acid residues. Several conformational transitions induced by ionic effects have been observed. These have been considered to indicate that the nucleosome is a dynamic structure capable of undergoing reversible conformational transitions; such conformational flexibility could be an important factor in chromatin transcription and replication (Chan and Piette, 1980).

3.4.2 Soluble Proteins

One of the first and most fruitful applications of spin labelling was the study of haemoglobin conformations. Advantage has been taken of the reactivity of the sulphhydryl group of cysteine at position 93 in the β-chain towards alkylating agents such as *N*-ethylmaleimide, iodoacetamide and *p*-chloromercuribenzoate. The spin labels used analogues of these reagents. Oxyhaemoglobin labelled with nitroxyl iodoacetamide gave composite ESR spectra showing a strongly immobilized component as well as a weakly immobilized one. The presence of a completely immobilized component in the spectrum of the spin labels bound to β-93 of oxyhaemoglobin turned out to be particularly advantageous. This is because a very small change in the protein structure in the region of the spin label will result in a significant change in the ESR spectrum. Thus at intermediate stages of deoxygenation of spin labelled oxy-haemoglobin both components of the ESR spectrum are present. Of special significance is the presence of isosbestic points in the low and high field regions of the superposition of spectra. These isosbestic points indicate the existence of two haemoglobin conformations and hence of only one conformational change associated with the oxygenation process. Moreover the results provide an opportunity to investigate directly the relationship between oxygenation and conformational change and to help discriminate between theories regarding allosteric transitions (Ogawa and McConnell, 1967).

The use of the iodoacetamide spin label has become a reference method for studying the structural transitions in haemoglobins, for example it has recently been applied to carp haemoglobin (Chien *et al.*, 1980).

Another soluble protein which has been extensively investigated by spin labelling is the bovine serum albumin. The fatty acid binding site of BSA has been studied using spin labelled derivatives of stearic acid and palmitic acid, while the steroid and indole binding sites have been studied using spin labelled derivatives of androstol and indole respectively (Morrisett *et al.*, 1975).

3.4.3 Spin Label Studies in Enzymology

It is possible to study enzyme kinetics by spin label ESR spectra where the reaction is accompanied by a change in the motional state of the nitroxides. Berliner and McConnell (1966) have acylated chymotrypsin at pH 4.5 with a *p*-nitrophenyl ester of a spin labelled carboxylic acid. The ESR spectrum of the strongly immobilized covalently bound label contains broad line components. When the pH is raised to 6.8, the spin labelled acyl group is released, giving a narrow line spectrum of the freely rotating nitroxide. The growth of the narrow line component is a sensitive indicator of deacylation and rate constants can be calculated. There was an agreement between the values of the rate constants calculated from ESR measurements of the release of spin labelled carboylic acid and from visible absorption measurements of the rate of release of *n*-nitrophenol. Similar kinetic experiments have been carried out on other systems (Morrisett, 1976).

The spectral changes occurring when a small nitroxide ligand binds (reversibly) to a macromolecule are associated, as already mentioned, with a shift to slower tumbling rates and consequently broader line shape spectra. These changes may be used for estimations of dissociation constants and binding stoichiometries of enzyme inhibitors. As an example to such an approach the work of Berliner and Wong (1975) with a spin labelled inhibitor for bovine galactosyl transferase could be mentioned.

There are several kinds of studies concerning the structure of enzymes that can be performed using spin labels.

(1) *Intramolecular distances* (5–25 Å) can be measured on the basis of interactions between the free electron and other nuclei, between the nitroxide radical and other paramagnetic centres or between the nitroxide and a fluorophore.

The spin lattice relaxation time of a nucleus is normally dominated by (magnetic) dipole–dipole interactions with surrounding magnetic dipoles (usually other nuclei). Since the electron magnetic moment exceeds that for a proton by 700-fold, a nearby paramagnet will have a profound effect on the nucleus of interest. As a consequence a 'relaxation enhancement' of the nucleus of interest will take place resulting in the shortening of its relaxation times. In a high resolution NMR spectrum the resonance of that nucleus will be broadened. From the shortening of the relaxation times and the broadenings in the NMR spectra the distances from the nitroxide to substrates bound at the active site can be calculated. Such studies have been performed on phosphofructokinase (Jones *et al.*, 1973), liver alcohol dehydrogenase (Mildvan and Weiner, 1969a,b), creatine kinase (Cohn *et al.*, 1971) and lysozyme (Wien *et al.*, 1972). Since these types of measurements imply the use of NMR techniques the reader is referred to the chapter of Krugh in the book of Berliner (1976).

Distances smaller than 10 Å between two different spin labels attached to the same enzyme can be measured on the basis of spin–spin exchange interactions and spin–spin dipolar interactions. The exchange interaction for nitroxide is characterized by a five line ESR spectrum, while dipolar interactions result in a complex, very broad triplet state spectrum. When a reagent carrying two spin labels (e.g. a biradical) is attached to the active site of an enzyme, the distance which separates the nitroxyl nitrogens may be goverened by the size of that site. Since only small changes in this distance can cause rather large changes in the ESR spectrum, this method has great potential for detecting subtle changes in active site conformation and geometry. However, because of difficulties in interpreting the complicated spectra of biradicals this approach has not yet seen wide use.

When a spin label and a paramagnetic ion are present in the same system, distances between the two can be measured based on a very large broadening of the spectral line shape due to the paramagnetic ion. The effect is an apparent 'quenching' of the ESR spectrum and the distance range over which occurs is 10–25 Å. This case is particularly applicable to metalloenzymes or enzymes utilizing metal ion cofactors in substrate catalysis, where a paramagnetic ion (e.g. Mn^{2+}) may be substituted for the natural metal (e.g. Mg^{2+} or Zn^{2+}). This approach has been used with creatine kinase (Taylor *et al.*, 1969) and phosphofructokinase (Jones *et al.*, 1973).

A fourth possibility is to measure distances between the spin label and a fluorophore. Fluorescence quenching is operative over a distance of 4–6 Å. One application has involved the spin labelled fatty acid induced quenching of intrinsic tryptophan fluorescence of albumin (Morrisett *et al.*, 1975).

(2) *Active site geometry* The size and shape of an enzyme active site and the relative positions of different groups within that site may be studied by an extension of the method of using spin labels as molecular rulers first applied by Hsia and Piette (1969) to measure the depth of antibody combining sites. The principle of the method is to measure the nitroxide mobility of a series of spin probes with increasing distances between the nitroxide moiety and the combining site.

For example, carbonic anhydrase has been studied with a series of sulphonamide spin probes of basic structure shown below. These are potent inhibitors of the action of this enzyme in hydrating CO_2 in the blood.

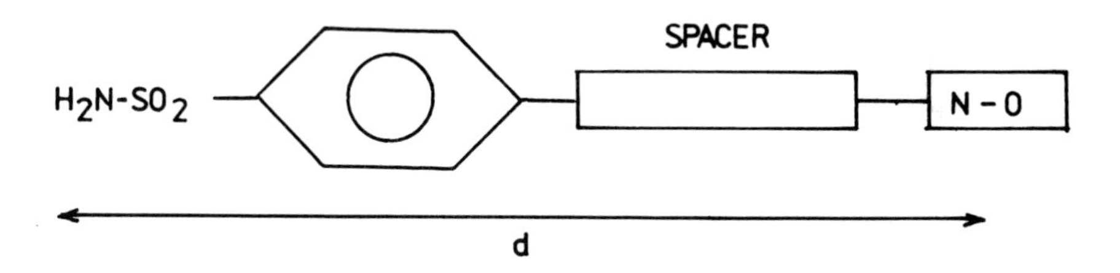

Spin probes with a small *d* are strongly immobilized. A monotonic increase in the mobility of the nitroxyl group with increasing *d* occurs until a sharp change in the mobility can be seen at a certain value of *d*. The experiments suggest that the active site of the enzyme is in a narrow cleft about 14.5 Å deep which is in excellent agreement with subsequent X-ray structure determinations. However there are pitfalls that one must consider when attempting such an approach on a new protein. If the spin probes are bound in a curved rather than extended conformation the estimated dimension *d* would appear longer than the actual value. Of more concern would be the case where the end of a spin probe, of length *d* greater than the depth of the combining site, became bound rigidly to some nearby hydrophobic site (secondary binding). This pitfall can be overcome by increasing the distance *d* in very small increments.

3.4.4 Spin Labelling of Model Membrane Systems

A variety of spin label studies have been performed on several types of model membrane systems. Some examples of the applications of spin labelling to liposomes and lipid–protein systems will be given here.

The average molecular motion of spin labels introduced into liposomal membranes reflects membrane fluidity. It has been found that the membrane interior can be in a highly fluid state with an apparent microviscosity comparable with that of a light oil. The membrane microviscosity may be estimated on the basis of spin label rotational correlation time. However, one must be aware of the fact that the lipid interior of a membrane is structurally different from that of a simple liquid, and the results obtained are somewhat dependent on the structure of the probe. For an isotropic liquid, fluidity is simply the reciprocal of viscosity. However, most membrane environments are not isotropic, and several correlation times may be required to describe the dynamic behaviour of a particular component. In addition there may be no correlation between the rates of molecular motion and the degree of molecular order in a system (Schreier *et al.*, 1978). It should be emphasized therefore that the absolute value of parameters derived from measurements on spin probes are not expected to be accurate.

Estimates of membrane fluidity can be made on the basis of measurements on whole spin labelled molecules introduced into liposomes. For example it was found that Tempo (I in Fig. 5) moves almost as freely in the hydrocarbon regions of membranes as it does in paraffin oil. A more detailed insight into the nature of the fluid lipid phase of membranes can be obtained from the motion of rigid steroid molecules intercalated in lipid bilayers. The ESR spectrum of steroid label VII (Fig. 5) which provides a model for the motion of cholesterol in membranes, has the important property that its splitting depends on the orientation and amplitude of motion of the spin label. It has been concluded that:

(1) the steroid orients with its long axis normal to the bilayer;
(2) the steroid molecule rotates rapidly about its long axis, and
(3) the steroid long axis performs an angular motion of limited amplitude.

Hence steroid molecules can form an integral part of the bilayer structure and the bilayer fluidity manifests itself in the long axis rotation and the limited angular motion. Schreier–Mucillo *et al.* (1973) found that the amplitude of steroid motion in egg phosphatidylcholine bilayers can be as large as 46° and this gradually decreases to a value of 28° on increasing the cholesterol content of the bilayers to an equimolar amount.

It is possible to examine the motion of the lipid chains themselves. The details of motion of the spin labelled fatty acids and phospholipids (IX and XI in Fig. 5 respectively) are interesting because these molecules are similar to natural components of membranes. For a spin labelled fatty acid two features of the molecule should be noted:

(1) the plane of the nitroxide ring is perpendicular to the long axis of the fatty acid chain, and
(2) the values of m and n, or the position of the nitroxide group on the hydrocarbon chain may be varied by synthesis.

For a series of spin labelled fatty acids in which the nitroxide group is at various positions between C-5 and C-16 of the fatty acid chain, the ESR spectra show increasing mobility as the distance between the carboxyl group and the position of the spin label increases (Hubbell and McConnell, 1969; Libertini *et al.*, 1969). This clearly corresponds to an increasing amplitude of motion of the hydrocarbon chains as one penetrates deeper into the hydrophobic interior of the bilayer. A similar gradient of motion was observed for a series of phosphatidylcholines (XI) with spin labelled fatty acids substituted for the β-acyl chain (Hubbell and McConnell, 1971).

One of the consequences of lipid chain fluidity is that phospholipids and steroid molecules are able to diffuse rapidly in the plane of the membrane. The first attempt to measure the rate of lateral diffusion in phospholipid bilayers was made by Kornberg and McConnell (1971a). They studied the *N*-methyl proton line broadening in phospholipid vesicles brought about by low concentrations of a spin labelled phosphatidylcholine. A more direct approach has been used by Devaux and McConnell (1972) to estimate the rate of this lateral diffusion process. If the spin label concentration is increased beyond the level normally used in spin labelling (1%) the spectrum is broadened by the so-called exchange interaction. As the broadening is determined by the rate of collision between spin label molecules it gives a measure of their rate of lateral diffusion. It was possible to estimate a diffusion constant D of 1.8×10^{-8} cm/sec at 25 °C for bilayers of phosphatidylcholine. The same method has been applied to natural membranes. For example the lateral diffusion constant of spin label molecules incorporated into sarcoplasmic reticulum membrane was $6 \times 10^{-8} cm^2/sec$ at 37 °C, a value very close to that obtained for the probe in an aqueous dispersion of a total membrane lipid extract (Scandella *et al.*, 1972). It seems that phospholipids diffuse in the plane of some biological membranes as rapidly as they do in artificial lipid membranes. For a value of $D = 10^{-8}$ cm^2/sec, each molecule exchanges with its neighbour about 10^7 times per second, so that lipid molecules are capable of redistributing rapidly over the surface of the membrane at a rate of the order of 1 μm per second. This means that a lipid molecule can travel the length of an *E. coli* bacterium in about 2 seconds.

In contrast to the freedom of lipid molecules to diffuse laterally in the plane of fluid bilayer structures, the movement of molecules from one leaflet of the bilayer to the other is a comparatively rare event. This is what one would expect taking into account the amphipathic properties of the lipid molecules. In order to achieve a transbilayer motion (flip-flop) the polar group must traverse the hydrophobic interior of the bilayer while at the same time exposing some of the hydrocarbon residues in water. Spin label studies provided one of the first methods of measuring the flip-flop rate of phospholipids in vesicle membranes. Kornberg and McConnell (1971b) prepared dispersions of phospholipids containing spin labelled phosphatidylcholine. Sodium ascorbate to which the membrane is impermeable reduces the label thus abolishing the paramagnetism in the external monolayer of the vesicle membrane without affecting the paramagnetism of internal facing molecules. The consequent asymmetry in the distribution of paramagnetic molecules between the two monolayers of the vesicle membrane decays with a half-time of 6.5 h at 30 °C. It was concluded that the probability of a spin labelled phosphatidylcholine molecule passing from the internal monolayer of the vesicle membrane to the external monolayer is 0.07/h at 30 °C. When the same technique was applied to biological membranes (McNamee and McConnell, 1973) the flip-flop rate of the spin label analogue was considerably faster than in vesicles of pure phospholipid suggesting that protein may influence the rate of transmigration.

Lipid hydrocarbon chains are not always in a fluid state. By increasing temperature, bilayers composed of a single phospholipid type undergo a well-defined phase transition from a rigid pseudocrystalline state to a form in which the chains are in the fluid state. The transition temperature (T_c) depends both on the hydrocarbon chains as well as on the headgroup of the phospholipids. At the T_c there is an abrupt change in the mobility of the hydrocarbon chains which can be detected as an abrupt variation of an ESR spectral parameter of a spin label introduced in the system, or as a discontinuity in the plot of ESR spectral parameters versus temperature.

Lipid bilayers composed of mixtures of phospholipids that have different transition temperatures do not undergo a single sharp phase transition, but a much broader transition in which different proportions of fluid and solid lipid can co exist in equilibrium over a certain range of temperatures. The evidence that complex membranes composed of lipids with varied structure may exhibit lateral phase separations is now considerable. The principal spin label method for measuring phase separations is based on the relative solubility of Tempo in membrane lipids and water as a function of temperature. The Tempo partitions between the aqueous and fluid lipid phases, as indicated in Fig. 11. When the lipid is in the 'solid' phase the Tempo is found only in the aqueous phase. As the amount of fluid-phase lipid increases the partitioning of Tempo into the lipid phase increases in proportion.

An important factor influencing the fluidity of lipid bilayer is the presence of protein. If a distribution of the spin label between phospholipid and protein occurs, the combination of spectra due to the spin label bound to the protein and free in the lipid phase may give an overall impression of immobilization, but this need not be due to the interaction between protein and phospholipids. Similar problems with regard to studies of protein–lipid interactions may occur with any probes (e.g. fluorescent probes) which can distribute themselves between protein and lipid.

The boundary layer lipid concept has been introduced from ESR studies of membrane proteins, especially those on cytochrome oxidase. Jost *et al.* (1973) gradually reduced the phospholipid content of purified cytochrome oxidase by successive acetone extractions and used spin labelled lipids to study protein–lipid interactions in the complex. At low lipid labels (below 0.2 mg phospholipid

/mg of protein) the spectrum corresponded to that of a highly immobilized nitroxide label, whereas at high lipid levels (0.5–0.7 mg phospholipid/mg protein) the label had a considerable degree of motional freedom. At intermediate lipid levels, two components of the spectrum appeared; a mobile component and an immobile component (or 'bound'). It has been concluded that the two components correspond to two classes of lipids:

(1) lipids tightly bound to the protein and immobilized (up to 0.2 mg of phospholipid/mg of protein), and
(2) the lipids in a fluid bilayer.

In recent years the boundary layer concept has been critically evaluated in various laboratories using spin labelling ESR. It has been stated that the immobilized lipids detected by ESR could reflect the existence of hydrocarbon chains trapped between adjacent proteins rather than a specific halo of phospholipids surrounding membrane proteins. Studies of Benga *et al.* (1979; 1981) on the cytochrome oxidse–lipid complex also suggest that identifying the residual lipid of the complex as tightly bound, immobilized lipid is a simplification, and that phospholipids in the vicinity of cytochrome oxidase may exhibit a high degree of mobility.

Studies by NMR indicate that lipid–protein interactions are not limited to a monomolecular annulus; on the contrary the protein is capable of influencing a much larger domain of the surrounding lipid. It should be pointed out that the time scale of NMR ($\sim 10^{-4}$ sec) is very different from the time scale of ESR ($\sim 10^{-9}$ sec). The immobile component of the ESR spectrum should be considered to correspond to the lipid that on the ESR time scale is influenced by the protein and not to the lipid tightly bound to protein.

3.4.5 *Studies on Natural Membranes*

Studies of membrane fluidity on bacterial membranes are useful since the lipid and fatty acid composition can be manipulated by changing the growth conditions. For example the moderately halophilic bacterium *Pseudomonas halosaccharolytica* grown at low temperatures and low salt concentration in the medium contain large amounts of unsaturated fatty acids in membrane lipids. As the growth temperature and salt concentration in the medium increase, the contents of saturated and cyclopropanoic fatty acids increase to more than half of the total fatty acids. In this condition the rotational correlation time of spin labelled fatty acids increases, indicating a decreased fluidity or a stiffening of the lipid bilayers.

Comparative studies of various membranes regarding the lipid composition and membrane fluidity as inferred by spin labelling shed further light on protein–lipid interactions in membranes. For example spin labelled fatty acids have been used to probe the fluidity of membrane lipids in human and rat liver mitochondria (Benga *et al.*, 1978). A greater mobility of spin label motion was noticed in the human membranes, a discontinuity of the electron spin resonance parameter as a function of temperature was noticed at a lower temperature for human than for rat. The unsaturation of lipids is lower in human than in rat liver mitochondria. Therefore the higher fluidity of human membrane can not be attributed to unsaturation, but might arise from a lesser immobilization of the lipids by protein in comparison with the rat liver membranes.

McConnell *et al.* (1972) devised as assay for determining the fraction of the lipid that is in a fluid state in a biological membrane. This assay is based on the extent of partitioning of Tempo into the membrane, measured as indicated in Fig. 11. It was found that 84% of the lipid in membranes of rabbit muscle sarcoplasmic reticulum is in a fluid state at 25 °C. However, it should be mentioned that the partition coefficient of Tempo in membrane lipids is small, so that it permits measurements only in the most concentrated membrane preparations. Wisniesky *et al.* (1974) introduced another spin label which partitions into membranes to a larger extent than Tempo and offers therefore more possibilities to characterize the 'fluidity' of cell membranes.

If proteins can affect the physical state of membrane lipids, much evidence has been provided that, in turn, lipid fluidity may influence the structure and function of membrane enzymes. Structural changes of the lipid phase may modify the configuration of membrane proteins, so that breaks in the enzyme activity appear at a temperature corresponding to a phase transition of the lipids.

The Ca-Mg-ATPase of sarcoplasmic reticulum is one of the membrane-bound enzymes most studied in this regard. Changes in the membrane structure around 20 °C as indicated by spin labelled fatty acids have been correlated with a change in the activation energy of Ca-Mg-ATPase occurring at the same temperature (Eletr and Inesi, 1972). This was interpreted in terms of a liquid-crystalline to crystalline phase transition. However, no phase transition is expected at such high temperature with the type of highly unsaturated lipids contained in sarcoplasmic reticulum. Alternative explanations have therefore been put forward. It was suggested that initiation of lipid segregation or cluster formation (Lee *et al.*, 1974) is more likely, or that changes in the apparent energy of activation of an enzyme process at particular temperatures does not reflect an event associated with the membrane lipids but rather is an intrinsic property of the protein itself.

Finally, attention should be paid to artefactual discontinuities in plots of spin label parameters as a function of temperature. Breaks in ESR spectral parameters may be artefactual due to problems in the measurement (see Schreier *et al.*, 1978, for a review).

Changes in membrane protein conformation may be detected using spin labels covalently bound to SH and NH_2 groups. The spin labels most commonly used are the nitroxide derivatives of maleimide (V in Fig. 5), iodoacetamide (VI in Fig. 5) and bromoacetamide.

An example of the use of the maleimide spin label is provided by the work of Nakamura *et al.* (1972) who studied the conformational changes induced by ATP on the ATPase of sarcoplasmic reticulum. The membrane labelled at pH 8.5 gave a composite spectrum with two classes of label-binding sites, one weakly and one strongly immobilized. The ratio of the concentrations of these two components changed upon addition of ATP to the sarcoplasmic reticulum in the presence of both Mg^{2+} and Ca^{2+}. When Mg^{2+} and Ca^{2+} were removed from the mixture by EDTA or EGTA, ATP caused no detectable change in the spectrum. The results were considered to indicate a conformational change of the ATPase in an intermediate stage of the reaction. Conformational changes in sarcoplasmic reticulum membrane correlated with Ca^{2+} transport have also been studied by Champeil *et al.* (1978, 1980) and Coan *et al.* (1979).

Other examples of applications of spin labels in studies on membranes may be found in the reviews of Gaffney and Lin (1976), Vignais and Devaux (1976), Marsh (1975), Gaffney and Chen (1977), and Schreier *et al.* (1978).

3.4.6 Biomedical Applications

Nitrophenyl derivatives have been used in a large number of studies on immunoglobulins (antibodies). The binding of spin labelled haptens (the small molecule antigen analogues) to antibodies and the homogenous analogues, the myeloma proteins, has been used both to measure binding constants and to obtain structural details of the binding state. Stryer and Griffith (1965) have shown that two moles of a spin labelled dinitrophenyl hapten were bound per mole of antibody and that the spin label was strongly immobilized in the binding state. Hsia and Piette (1969) extended these studies by using dinitrophenyl–nitroxides in which the distance between the dinitrophenyl and nitroxide groups varied. As the distance increased, the nitroxide moiety of the hapten bound to antidinitrophenyl antibodies became more mobile. Since the mobility increased very rapidly as the dinitrophenyl–nitroxide distance became greater than 11 Å, an average depth of 10 Å was estimated for the antibody–hapten combining site. This approach has been further extended by Dwek *et al.* (1975) to the analysis of the dinitrophenyl hapten spin labels combining sites of a myeloma protein. Using spin labels of increasing length and with six-membered rings or five-membered rings it was possible to calculate the length (11–12 Å) and lateral dimensions (9 Å $\times$ 6 Å) of the antibody combining site.

A simple, sensitive immunoassay method based on the spin labelling ESR has been developed and called spin membrane immunoassay or SMIA (Hsia and Tan, 1978). The intensity of the ESR spectrum of a spin label is proportional to its concentration, except at high concentrations, where there is spin exchange interaction between the unpaired electrons and a strong broadening of the ESR signal. This effect provides a sensitive and novel technique for monitoring the lysis of liposomes or erythrocyte ghosts containing a concentrated solution of a water-soluble, membrane-impermeable spin label such as Tempo-choline chloride: the lysis results in a great enhancement of the ESR signal intensity. For a long time complement-mediated immune lysis of erythrocytes or sensitized erythrocytes (haemolysis) has been used as an assay technique to determine concentrations of antigens, antibodies or complement. Humphries and McConnell (1974) showed that sensitized liposomes or erythrocyte ghosts loaded with Tempo-choline chloride can be used for measuring the complement-mediated immune lysis. SMIA has certain advantages in sensitivity and in convenience. Since the ESR signal is proportional to the extent of label release, optically opaque 20 μl samples may be monitored. Prior purification of the samples is not required. SMIA has also advantages compared to radioimmunoassay where one of the major limitations is the need to separate free and antibody-bound labels (by precipitation, partition chromatography or ultracentrifugation). All these procedures are time consuming, relatively expensive, and increase the possibility of error. Further limiting factors of radioimmunoassay are the hazard, short shelf-life for some isotopes, and cost of radioactive materials. SMIA does not require separation procedures, involves no radiation hazards, and has the added advantages of speed, simplicity, and small sample volume. There is the possibility of automation with great potential for performing multiple assays because different spin labels and sensitizers can simultaneously be used to detect different antigens.

Spin labelling has been used in studies of the molecular mechanisms by which some genetic diseases arise. Erythrocyte membranes have been studied in a series of neurological disorders that are suggested to be associated with generalized membrane defects (see Butterfield, 1977, and Butterfield and Markesberry, 1980, for reviews). These include Huntington's disease, Friedreich's ataxia, Alzheimer's disease, amyotrophic lateral sclerosis, and myotonic and Duchenne muscular dystrophy.

3.5 Recent Developments in the Biological Applications of ESR Spectroscopy

As previously discussed, conventional ESR spectroscopy of nitroxide radicals is limited to the measurement of correlation times shorter (faster) than 10^{-7} sec.

As motion becomes slower than 10^{-7} sec, the effects on the ESR spectra become less and less as the spectra asymptotically approach the line shape expected from a rigid form, and the ability to use spin labels is limited by the signal-to-noise ratio, so that very slow motion cannot be studied by conventional ESR.

However, an ESR technique referred to as saturation transfer spectroscopy (ST-ESR) has been developed which has maximum sensitivity to slow rotational motions (correlation times $10^{-7} < \tau < 10^{-3}$ sec). ST-ESR has been developed by J. Hyde, L. Dalton and D. Thomas (Hyde and Dalton, 1972; Hyde and Thomas, 1973) and theoretical aspects are discussed by Thomas *et al.* (1976).

ST-ESR has proven applicable to the study of soluble proteins such as spin labelled haemoglobin (Hyde and Thomas, 1973) and spin labelled subfragment-1 in myosin (Thomas *et al.*, 1975), and has been used in studies on spin labelled tobacco mosaic virus protein (Hemminga *et al.*, 1977). Since membrane-bound proteins experience even slower motions than soluble proteins, ST-ESR seems to be a useful technique to investigate their properties. By this technique Devaux and co-workers (Baroin *et al.*, 1977; Rousselet and Devaux, 1977) have found it possible to differentiate between proteins 'floating' in a fluid lipid bilayer and proteins tightly anchored or so tightly packed that no motion occurs. For example membrane-bound rhodopsin has a rotational correlation time $\tau = 20\,\mu$sec at 20 °C, while the cholinergic receptor in *Torpedo* membrane fragments or the ADP carrier in mitochondria seems to be immobilized ($\tau \approx 10$ msec) possibly because of strong protein–protein interactions.

A mention should be made of a new technique which has recently been applied to free radical studies in biological systems, that of spin trapping. This technique involves the use of a compound which forms a stable free radical by reacting covalently with an unstable radical. In this way one can use ESR to study radicals of biological interest which are highly reactive and never reach a concentration high enough to be otherwise detected. Nitrones and nitroso compounds are the spin traps most commonly used and the technique has allowed investigations of a variety of systems; microsomal lipid peroxidation, photolysis of chlorophylls, production of radicals formed during the liver metabolism of drugs and carcinogenic nitrosamines. Biological applications of spin trapping have been reviewed by Janzen (1980), McCay *et al.* (1980) and Finkelstein *et al.* (1980).

3.6 Advantages and Disadvantages of Spin Labelling

In any reporter group technique the introduction of the label perturbs the system as a result of its size and chemical properties. This also applies to spin labelling. It is therefore clear that the labels or probes cannot be expected to yield information that is in exact quantitative agreement with physical measurements in the absence of labels. The use of probes is probably best suited to measuring differences between samples. In this regard, measurements on membranes with spin labelled fatty acids and phospholipids and with Tempo, are similar in many cases to measurements with fluorescent probes.

ESR is less sensitive than fluorescence and considerably more sensitive than nuclear magnetic resonance. Although for very rapidly tumbling nitroxides, as few as 10^{-13} mole can be detected, the practical upper limit is 10^{-10} to 10^{-11}

mole. If the nitroxide is more immobilized the usual range of bulk concentrations that are easy to work with are 10^{-5} to 10^{-6} M.

The fact that the sample can be optically opaque is an important advantage of ESR.

ESR spectra contain information about molecular motion, orientation and the nature of chemical environments. It has been stated that no other technique allows such a powerful combination of flexibility of approach, sensitivity to concentration and conformation, and interpretability of results.

Appendix

Commercial Sources of Spin Labelling Equipment and Supplies

ESR spectrometers	
Varian Associates, USA	611 Hansen Way, Palo Alto, California 94303, USA
JEOL, Japan	1418 Nakagami, Akishima, Tokyo 196
Brucker ER tt, FRG	Brucker Analitische Messtechnik, Silberstriefen B 7512, Rheinstetten 4-F_6, FRG
ART – 6, Roumania	Institute of Physics & Nuclear Engineering, R-76900 Magurele-Bucharest, Roumania
Quartz sample cells and glassware	
Varian Instruments	611 Hansen Way, Palo Alto, California 94303, USA
Wilmad Glass Company Inc.	U.S. Route 40 and Oak Road, Buena, New Jersey 08310, USA
James F. Scanlon Company	2428 Baseline Avenue, Solvang, California 93463, USA
Spin labels and nitroxide precursors	
SYVA	3221 Porter Drive, Palo Alto, California 94304, USA
Aldrich Chemical Company Inc.	940 W. St. Paul Avenue, Milwaukee, Wisconsin 53233, USA
Eastman Organic Chemicals	Eastman Kodak Company, 343 State Street, Rochester, New York, USA
Frinton Labs	P.O. Box 301, Grant Avenue, So. Vineland, New Jersey 08360, USA

The listing is by no means intended to be complete

References

Baroin, A., Thomas, D. D., Osborne, B., and Devaux, P. F. (1977) Saturation transfer electron paramagnetic resonance on membrane-bound proteins. I. Rotational diffusion of rhodopsin in the visual receptor membrane. *Biochem. Biophys. Res. Commun.* **78**, 442–447.

Bauer, R. S., and Berliner, L. J. (1979) Spin label investigations of α-chymotrypsin active site structure in single crystals. *J. Mol. Biol.* **128**, 1–19.

Benga, G., Hodarnau, A., Bohm, B., Borza, V., Tilinca, R., Dancea, S., Petruscu, I., and Ferdinand, W. (1978) Human liver mitochondria: relation of a particular lipid composition to the mobility of spin-labelled lipids. *Eur. J. Biochem.* **84**, 625–633.

Benga, G., Popescu, O., and Pop, V. (1979) Protein–lipid interactions in biological membranes. Cytochrome oxidase–lipid complex: Spin label studies. *Rev. roum. Biochim.* **16**, 175–181

Benga, G., Porumb, T., and Wrigglesworth, J. M. (1981) Estimation of lipid regions in a cytochrome oxidase–lipid complex using spin labelling electron spin resonance: Distribution effects on the spin label. *J. Bioenergetics Biomembranes* **13**, 269–283

Benga, G., and Strach, S. J. (1975) Interpretation of the electron spin resonance spectra of nitroxide-maleimide-labelled proteins and the use of this technique in the study of albumin and biomembranes. *Biochim. Biophys. Acta* **400**, 69–79

Berliner, L. J. (ed.) (1976) *Spin Labelling, Theory and Applications,* Vol. 1. Academic Press, New York

Berliner, L. J. (ed.) (1979) *Spin Labelling, Theory and Applications,* Vol. 2. Academic Press, New York

Berliner, L. J. (1981) Using the spin label method in enzymology. In: *Spectroscopy in Biochemistry* (ed. J. E. Bell), pp. 1–56. CRC Press, Boca Raton, Florida

Berliner, L. J., and McConnell, H. M. (1966) A spin-labelled substrate for α-chymotrypsin. *Proc. Nat. Acad. Sci. USA* **55**, 708–712

Berliner, E. J., and Wong, S. S. (1975) Manganese (II) and spin-labelled uridine 5′-diphosphate binding to bovine galactosyltransferase. *Biochemistry* **14**, 4977–4982

Bobst, A. M., Hakam, A., Langemeier, P. W., and Kouidou, S. (1979) Electron spin resonance melting of chemically spin labelled nucleic acids. *Arch. Biochem. Biophys.* **194**, 171–178

Bolton, J. R., Borg, D. C., and Swartz, H. M. (1972) Experimental aspects of biological electron spin resonance studies. In: *Biological Applications of Electron Spin Resonance* (eds H. M. Swartz, J. R. Bolton, and D. C. Borg), pp. 483–539. Wiley (Interscience), New York

Burr, M., and Koshland, D. E. Jr. (1964) Use of reporter groups in structure-function studies of proteins. *Proc. Nat. Acad. Sci. USA* **52**, 1017–1024

Butterfield, D. A. (1977) Electron spin resonance studies of erythrocyte membranes in muscular dystrophy. *Acc. Chem. Res.* **10**, 111–116

Butterfield, D. A., and Markesberry, W. R. (1980) Specificity of biophysical and biochemical alterations in erythrocyte membranes in neurological disorders. Huntington's disease, Friedreich's ataxia, Alzheimer's disease, amyotrophic lateral sclerosis and myotonic and Duchenne muscular dystrophy. *J. Neurol. Sci.* **47**, 261–271

Calvin, M., Wang, H. H., Entine, D., Gill, P., Ferruti, P., Harpold, M. A., and Klein, M. P. (1969) Biradical spin labelling for nerve membranes. *Proc. Nat. Acad. Sci. USA* **63**, 1–8

Cannon, B., Polnaszek, C. F., Butler, K. W., Eriksson, L. E. G., and Smith, I. C. P. (1975) The fluidity and organization of mitochondrial membrane lipids of the brown adipose tissue of cold-adapted rats and hamsters as determined by nitroxide spin probes. *Arch. Biochem. Biophys.* **167**, 505–518

Champeil, P., Büschlen-Boucly, S., Bastide, F., and Gary-Bobo, S. (1978) Sarcoplasmic reticulum ATPase. Spin labelling detection of ligand-induced changes in the relative reactivities of certain sulfhydryl groups. *J. Biol. Chem.* **253**, 1179–1186

Champeil, P., Rigaud, J. L., and Gary-Bobo, C. M. (1980) Calcium translocation mechanism in sarcoplasmic reticulum vesicles, deduced from location studies of protein-bound spin labels. *Proc. Nat. Acad. Sci. USA* **77**, 2405–2409

Chan, D. C. F., and Piette, L. H. (1980) ESR spin label studies of the nucleosome core particle and histone core. *Biochim. Biophys. Acta* **623**, 32–45
Chapman, D., and Wallach, D. F. H. (1973) *Biological Membranes. Physical Fact and Function,* Vol. 2. Academic Press, London
Chien, J. C. W., Dickinson, L. C., Snyder, F. W. Jr., and Mayo, K. H. (1980) Circular dichroism and spin-label studies of carp haemoglobin. *J. Mol. Biol.* **142**, 75–91
Coan, C., Verjovski-Almeida, S., and Inesi, G. (1979) Ca^{2+} regulation of conformational states in the transport cycle of spin-labelled sarcoplasmic reticulum ATPase. *J. Biol. Chem.* **254**, 2968–2974
Cohn, M., Diefenbach, H., and Taylor, J. S. (1971) Magnetic resonance studies of the interaction of spin-labelled creatine kinase with paramagnetic-substrate complexes. *J. Biol. Chem.* **246**, 6037–6042
Devaux, P. F., and McConnell, H. M. (1972) Lateral diffusion in spin-labelled phosphatidylcholine multilayers. *J. Am. Chem. Soc.* **94**, 4475–4481
Dwek, R. A., Knott, J. C., Marsh, D., McLaughlin, A. C., Press, E. M., Price, N. C., and White, A. I. (1975) Structural studies on the combining site of the myeloma protein MOPC 315. *Eur. J. Biochem.* **53**, 25–39
Eletr, S., and Inesi, G. (1972) Phase changes in the lipid moieties of sarcoplasmic reticulum membranes induced by temperature and protein conformational changes. *Biochim. Biophys. Acta* **290**, 178–185
Finkelstein, E., Rosen, G. M., and Rauckman, E. J. (1980) Spin trapping of superoxide and hydroxyl radical: Practical aspects. *Arch. Biochem. Biophys.* **200**, 1–16
Forrester, A. R., Hay, J. M., and Thomson, R. H. (1968) *Organic Chemistry of Stable Free Radicals.* Academic Press, New York
Gaffney, B. J., and Chen, S. C. (1977) Spin label studies of membranes. In: *Methods in Membrane Biology* (ed. E. Korn), Vol. 8, pp. 291–357. Plenum Press, New York
Gaffney, B. J., and Lin, D. C. (1976) Spin-label measurements of membrane-bound enzymes. In: *The Enzymes of Biological Membranes* (ed. A. Martonosi), Vol. 1, pp. 71–90. Plenum Press, New York
Giotta, G. J., and Wang, H. H. (1973) Mobility of spin-labelled sulfhydryl sites in excitable tissue. *Biochim. Biophys. Acta* **298**, 986–994
Goldman, S. A., Bruno, G. V., and Freed, J. H. (1972) Estimating slow motional rotational correlation times for nitroxides by electron spin resonance. *J. Phys. Chem.* **76**, 1858–1860
Griffith, O. H., and McConnell, H. M. (1966) A nitroxide maleimide spin label. *Proc. Nat. Acad. Sci. USA* **55**, 8–11
Griffith, O. H., and Waggoner, A. S. (1969) Nitroxide free radicals: Spin labels for probing molecular structure. *Acc. Chem. Res.* **2**, 17–24
Hamilton, C. L., and McConnell, H. M. (1968) Spin labels. In: *Structural Chemistry and Molecular Biology* (eds A. Rich and N. Davidson), pp. 115–149. W. H. Freeman, San Francisco
Hemminga, M. A., De Jager, P. A., and De Wit, J. L. (1977) Saturation transfer electron paramagnetic resonance spectroscopy of spin labelled tobacco mosaic virus protein. *Biochem. Biophys. Res. Commun.* **79**, 635–639
Hoffman, B. M., Schofield, P., and Rich, A. (1969) Spin-labelled transfer RNA. *Proc. Nat. Acad. Sci. USA* **62**, 1195–1202
Hsia, J. C., and Piette, L. H. (1969) Spin-labelling as a general method in studying antibody active site. *Arch. Biochem. Biophys.* **129**, 296–307
Hsia, J. C., and Tan, C. T. (1978) Membrane immunoassay: Principle and application of spin membrane immunoassay. *Ann. N.Y. Acad. Sci.* **308**, 139–147
Hubbell, W. L., and McConnell, H. M. (1969) Orientation and motion of amphiphilic spin labels in membranes. *Proc. Nat. Acad. Sci. USA* **64**, 20–27

Hubbell, W. L., and McConnell, H. M. (1971) Molecular motion in spin-labelled phospholipids and membranes. *J. Am. Chem. Soc.* **93**, 314–326

Humphries, G. K., and McConnell, H. M. (1974) Immune lysis of liposomes and erythrocyte ghosts loaded with spin label. *Proc. Nat. Acad. Sci. USA* **71**, 1691–1694

Hyde, J. S., and Dalton, L. (1972) Very slowly tumbling spin labels: adiabatic rapid passage. *Chem. Phys. Lett.* **16**, 568–572

Hyde, J. S., and Thomas, D. D. (1973) New EPR methods for the study of very slow motion: Application to spin-labelled hemoglobin. *Ann. N.Y. Acad. Sci.* **222**, 680–692

Janzen, E. G. (1980) A critical review of spin trapping in biological systems. In: *Free Radicals in Biology* (ed. W. A. Pryor), Vol. IV, pp. 115–153. Academic Press, New York

Jones, G. L., and Woodbury, D. M. (1978) Reappraisal of the electron spin resonance spectra of maleimide and iodoacetamide spin labels in erythrocyte ghosts. *Arch. Biochem. Biophys.* **190**, 611–616

Jones, R., Dwek, R. A., and Walker, I. O. (1973) Spin-labelled phosphofructokinase and its interactions with ATP and metal-ATP complexes as studied by magnetic-resonance methods. *Eur. J. Biochem.* **34**, 28–40

Jost, P. C., and Griffith, O. H. (1972) Electron spin resonance and the spin labelling method. In: *Methods in Pharmacology* (ed. C. Chignell), Vol. 2, pp. 223–276. Appleton-Century-Crofts

Jost, P. C., Griffith, O. H., Capaldi, R. A., and Vanderkooi, G. (1973) Evidence for boundary lipid in membranes. *Proc. Nat. Acad. Sci. USA* **70**, 480–484

Jost, P. C., Waggoner, A. S., and Griffith, O. H. (1971) Spin labelling and membrane structure. In: *Structure and Function of Biological Membranes* (ed. L. Rothfield), pp. 84–144. Academic Press, New York

Kawamura, T., Matsunami, S., and Yonezawa, T. (1967) Solvent effects on the g-value of di-*t* butyl nitric oxide. *Bull. Chem. Soc. Japan* **40**, 1111–1115

Keith, A., Waggoner, A., and Griffith, O. H. (1968) Spin-labelled mitochondrial lipids in Neurospora crassa. *Proc. Nat. Acad. Sci. USA* **61**, 819–826

Kivelson, D. (1960) Theory of esr linewidths of free radicals. *J. Chem. Phys.* **33**, 1094–1106

Kornberg, R. D., and McConnell, H. M. (1971a) Lateral diffusion of phospholipids in a vesicle membrane. *Proc. Nat. Acad. Sci. USA* **68**, 2564–2568

Kornberg, R. D., and McConnell, H. M. (1971b) Inside-out transition of phospholipids in vesicle membranes. *Biochemistry* **10**, 1111–1120

Landsberger, F. R., Paxton, J., and Lenard, J. (1971) A study of intact human erythrocytes and their ghosts using stearic acid spin labels. *Biochim. Biophys. Acta* **266**, 1–6

Laporte, A., Richard, H., Bonnaud, E., Henry, P., Vital, A., and Georgescauld, D. (1979) A spin label study of myelin fluidity with normal and pathological peripheral nerves. *J. Neurol. Sci.* **43**, 345–356

Lawrence, J. J., Berne, L., Ouvrier-Buffet, J. L., and Piette, L. H. (1980) Spin label studies of histone H1–DNA interaction. *Eur. J. Biochem.* **107**, 263–269

Lee, A. G., Birdsall, N. M. J., Metcalfe, J. C., Toon, P. A., and Warren, G. B. (1974) Clusters in lipid bilayers and the interpretation of thermal effects in biological membranes. *Biochemistry* **13**, 3699–3705

Libertini, L. J., Waggoner, A. S., Jost, P. C., and Griffith, O. H. (1969) Orientation of lipid spin labels in lecithin multilayers. *Proc. Nat. Acad. Sci. USA* **64**, 13–19

Likhtenstein, G. T. (1974) *Spin Labelling.* Nauka Publishing House, Moscow

Marsh, D. (1975) Spectroscopic studies of membrane structure. In: *Essays in Biochemistry* (eds P. Campbell and W. N. Aldridge), Vol. 11, pp. 139–180. Academic Press, London

McCay, P. B., Noguchi, T., Fong, K. L., Lai, E. K., and Poyer, J. L. (1980) Production of radicals from enzyme systems and the use of spin traps. In: *Free Radicals in Biology* (ed. W. A. Pryor), Vol. IV, pp. 155–186. Academic Press, New York

McConnell, H. M., and McFarland, B. G. (1970) Physics and chemistry of spin labels. *Quart. Rev. Biophys.* **3**, 91–136

McConnell, H. M., Wright, K. L., and McFarland, B. G. (1972) The fraction of the lipid in a biological membrane that is in a fluid state: a spin label assay. *Biochem. Biophys. Res. Commun.* **47**, 273–281

McNamee, M. G., and McConnell, H. M. (1973) Transmembrane potentials and phospholipid flip-flop in excitable membrane vesicles. *Biochemistry* **12**, 2951–2958

Mildvan, A. S., and Weiner, H. (1969a) Interaction of a spin-labelled analogue of nicotinamide-adenine nucleotide with alcohol dehydrogenase. *J. Biol. Chem.* **244**, 2465–2475

Mildvan, A. S., and Weiner, H. (1969b) Interaction of a spin-labelled analogue of nicotinamide-adenine nucleotide with alcohol dehydrogenase. II. Proton relaxation rate and electron paramagnetic resonance studies of binary and ternary complexes. *Biochemistry* **8**, 552–561

Morrisett, J. D. (1976) The use of spin labels for studying the structure and function of enzymes. In: *Spin Labelling. Theory and Applications.* (ed. L. J. Berliner), Vol. 1. Academic Press, New York

Morrisett, J. D., Pownall, H. J., and Gotto, A. M. Jr. (1975) Bovine serum albumin. Study of the fatty acid and steroid binding sites using spin-labeled lipids. *J. Biol. Chem.* **250**, 2487–2494

Nakamura, H., Hori, H., and Mitsui, T. (1972) Conformational change in sarcoplasmic reticulum induced by ATP in the presence of magnesium ion and calcium ion. *J. Chem.* **72**, 635–642

Ogawa, S., and McConnell, H. M. (1967) Spin-label study of hemoglobin conformations in solution. *Proc. Nat. Acad. Sci. USA* **58**, 19–26

Ohnishi, S., and McConnell, H. M. (1965) Interaction of the radical ion of chlorpromazine with deoxyribonucleic acid. *J. Am. Chem. Soc.* **87**, 2293

Rottem, S., Hubbell, W. L., Hayflick, L., and McConnell, H. M. (1970) Motion of fatty acid spin labels in the plasma membrane of Mycoplasma. *Biochim. Biophys. Acta* **219**, 104–113

Rousselet, A., and Devaux, P. F. (1977) Saturation transfer electron paramagnetic resonance on membrane bound proteins. II. Absence of rotational diffusion of the cholinergic receptor protein in *Torpedo marmorata* membrane fragments. *Biochem. Biophys. Res. Commun.* **78**, 448–454

Rozantsev, E. G., and Sholle, V. D. (1971) Synthesis and reactions of stable nitroxyl radicals. I. Synthesis. *Synthesis* pp. 190–250

Scandella, Ç. J., Devaux, P. F., and McConnell, H. M. (1972) Rapid lateral diffusion of phospholipids in rabbit sarcoplasmic reticulum. *Proc. Nat. Acad. Sci. USA* **69**, 2056–2060

Schreier, S., Polnaszek, C. F., and Smith, I. C. P. (1978) Spin labels in membranes. Problems in practice. *Biochim. Biophys. Acta* **515**, 395–436

Schreier-Muccillo, S., Marsh, D., Dugas, D., Schneider, H., and Smith, I. C. P. (1973) A spin probe study of the influence of cholesterol on motion and orientation of phospholipids in oriented multibilayers and vesicles. *Chem. Phys. Lipids* **10**, 11–27

Shimshick, E. J., and McConnell, H. M. (1973) Lateral phase separations in phospholipid membranes. *Biochemistry* **12**, 2351–2360

Sholle, V. D., Kagan, E. S., Michailov, V. J., Rozantsev, E. G., Frangopol, P. T., Frangopol, M., Pop, V. I., and Benga, G. (1980) A new spin label for SH groups in proteins: The synthesis and some applications in labeling of albumin and biomembranes. *Rev. roum. Biochim.* **17**, 291–298

Smith, I. C. P. (1972) The spin label method. In: *Biological Applications of Electron Spin Resonance* (eds H. M. Swartz, J. R. Bolton and D. C. Borg), pp. 483–539. Wiley (Interscience), New York

Smith, I. C. P., Schreier-Mucillo, S., and Marsh, D. (1976) Spin labeling. In: *Free Radicals in Biology* (ed. W. Pryor), Vol. 1, pp. 149–197. Academic Press, New York

Stanacev, N., Stuhne-Sekalec, L., Schreier-Muccillo, S., and Smith, I. C. P. (1972) Biosynthesis of spin-labeled phospholipids. Enzymatic incorporation of spin-labeled stearic acid into phosphatidic acid. *Biochem. Biophys. Res. Commun.* **46**, 114–119

Stone, T. J., Buckman, T., Nordio, P. L., and McConnell, H. M. (1965) Spin labelled biomolecules. *Proc. Nat. Acad. Sci. USA* **54**, 1010–1017

Stryer, L., and Griffith, O. H. (1965) A spin labelled hapten. *Proc. Nat. Acad. Sci. USA* **54**, 1785–1791

Swartz, H. M., Bolton, J. R., and Borg, D. C. (1972) *Biological Applications of Electron Spin Resonance.* Wiley (Interscience) New York

Taylor, J. S., Leigh, J. S., and Cohn, M. (1969) Magnetic resonance studies of spin labelled creatine kinase system and interaction of two paramagnetic probes. *Proc. Nat. Acad. Sci. USA* **64**, 219–226

Thomas, D. D., Dalton, L. R., and Hyde, J. S. (1976) Rotational diffusion studied by passage saturation transfer electron paramagnetic resonance. *J. Chem. Phys.* **65**, 3006–3024

Thomas, D. D., Seidel, J. C., Hyde, J. S., and Gergely, J. (1975) Motion of subfragment-1 in myosin and its supramolecular complexes: Saturation transfer electron paramagnetic resonance. *Proc. Nat. Acad. Sci. USA* **72**, 1729–1733

Tourtellotte, M., Branton, D., and Keith, A. (1970) Membrane structure: Spin labelling and freeze-etching of *Mycoplasma laidlawii. Proc. Nat. Acad. Sci. USA* **66**, 909–916

Vignais, P. M., and Devaux, P. F. (1976) The use of spin labels to study membrane-bound enzymes, receptors and transport systems. In: *The Enzymes of Biological Membranes* (ed. A. Martonosi), Vol. 1, pp. 91–117. Plenum Press, New York

Viret, J., Leterrier, F., and Bourre, J. M. (1979) A spin label study of sciatic nerves from quaking, jumpy and trembler mice. *Biochim. Biophys. Acta* **558**, 141–146

Wien, R. W., Morrisett, J. D., and McConnell, J. M. (1972) Spin label-induced nuclear relaxation. Distances between bound saccharides, histidine-15 and tryptophan-123 on lysozyme in solution. *Biochemistry* **11**, 3707–3716

Williams, B. L., and Wilson, K. (1979) *A Biologist's Guide to Principles and Techniques of Practical Biochemistry*, Edward Arnold, London

Wisnieski, B. J., Parkes, J. G., Huang, Y. O., and Fox, C. F. (1974) Physical and physiological evidence for two phase transitions of cytoplasmic membranes of animal cells. *Proc. Nat. Acad. Sci. USA* **71**, 4381–4385

Biochemical Research Techniques
Edited by J. M. Wrigglesworth

4
Principles and Applications of High Performance Liquid Chromatography

ROBERT F. G. BOOTH and PETER J. QUINN

Department of Biochemistry, Chelsea College, University of London, London SW3 6LX

4.1 Introduction

A wide variety of physical separation methods based upon the sample partioning between a moving phase, which can be a gas or a liquid, and a stationary phase, which may be either a liquid or a solid have been developed for molecules of biological interest. Liquid chromatography is generally regarded as the forerunner of all modern chromatographic methods but, until recent developments in instrumentation and column packing materials, it remained a comparatively little used technique. High performance liquid chromatography (HPLC) is now becoming recognized as one of the most powerful analytical and preparative techniques, the applications of which appear to be limited only by the dedication and skill of the operator. The major advantages of the method include high resolution, speed, sensitivity, automatic operation and an unsurpassed range of applications particularly in the pharmaceutical and biomedical areas.

In this chapter an introduction to the theory of the HPLC method will be provided together with an outline of different chromatographic systems that are available. Finally, some applications of the method to problems of biochemical separations will be described with particular emphasis on the uses of novel sample detection systems.

4.2 Principles of HPLC

4.2.1 Chromatographic Parameters

The objective of any chromatography system is to separate components of a mixture. Liquid chromatography systems exploit different affinities of the components for a stationary phase packed into a column and a mobile liquid phase that passes through it. Successful separation can be determined by monitoring the effluent from the column, using a suitable sample detector, and an elution chromatogram usually relates the concentration of the components in the mobile phase with time from application of the mixture to the column. A typical chromatogram is illustrated in Fig. 1 which shows the separation of components A and B applied to the column as a mixture at time $t = 0$. It can be seen that components A and B are retained on the column for different times, $t_r = 10$ min and $t_r = 13$ min respectively. This is due to the fact that B has a greater affinity than A for the stationary phase in the column and it represents the basis upon which the separation is achieved.

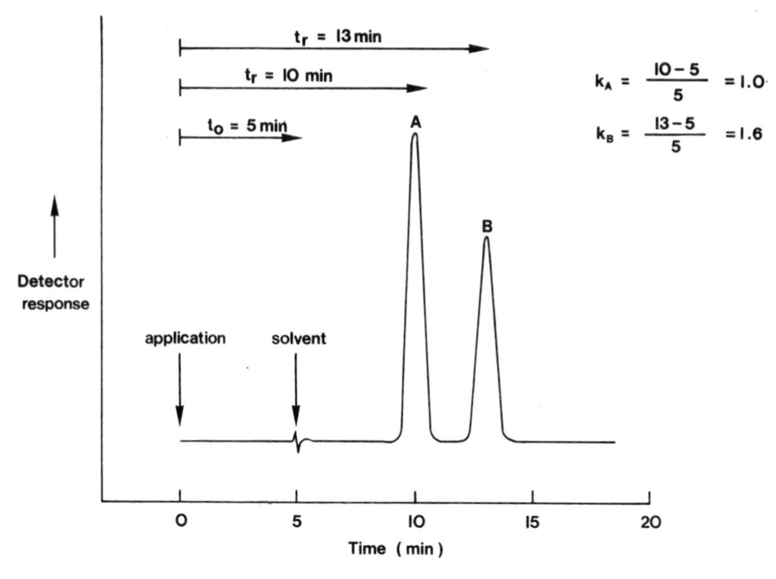

Fig. 1 A chromatogram showing separation of two components, A and B, applied to a column as a mixture at time, $t = 0$

A useful method of explaining this parameter is the column capacity ratio, usually denoted k, which is the ratio of the retention time $(t_r - t_0)$ of a solute component to the retention time of unretained material t_0 assuming that the rate of flow of solvent through the column is constant.

$$k = (t_r - t_0)/t_0 \tag{1}$$

If the assumption is valid then the ratio can be more correctly, but less conveniently, expressed in terms of retention volumes. As expressed in Eq. 1 the capacity ratio is a measure of the ratio of the time spent by the solute in the stationary phase to the time spent in the mobile phase. Small values of k indicate that the components are not significantly retained by the column and consequently there is poor separation of solute mixtures.

The efficiency of a chromatographic column is measured by the number of theoretical plates, N, to which the column is equivalent. According to this concept which was originally formulated to describe the process of distillation, the chromatographic column is considered to consist of a series of hypothetical layers or plates in which the solute concentrations in the relevant phases are assumed to be equilibrated. The value of N is a useful measure of the extent of spreading of a solute band as it passes through the column and this can be judged from the profile of the peak eluted from the column. More efficient columns produce narrow peaks of solute elution and the spreading of the peak is correlated with decreasing values of N. The number of theoretical plates can be calculated from Eq. 2:

$$N = 16(t_r/W)^2 \tag{2}$$

where W is the width of the base of the peak, in the same units as t, measured by extrapolation of tangents at the points of inflection to the baseline. In general the value of N is independent of the retention time of a solute but is proportional to the column length. To enable comparisons to be made between columns of different length the preferred measure of column efficiency is the height equivalent to a theoretical plate (H) which is related to column length, L, by Eq. 3:

$$H = L/N \tag{3}$$

Efficient columns are thereby characterized by small values of H under a given set of operating conditions. In general, values of H are less for small particles through which the mobile phase flows slowly, for mobile phases that are less viscous (higher separation temperatures), and for small sample molecules. Factors associated with a decrease in the efficiency of chromatographic columns include irregular flow of the mobile phase, longitudinal diffusion of the solute within the column and deviations from sorption equilibrium giving rise to variations in mass transfer between the two phases. Uneven flow through columns is caused by the creation of channels within the bed which represent regions of low resistance to solvent flow and aggregates of bed material that give rise to eddy diffusion. Such effects can be reduced by using particles of uniform size distribution in the column and packing them evenly throughout the length of the column. The rate of molecular diffusion in liquids is about five orders of magnitude less than in the gas phase and provided retention times of solutes on columns are not excessively long, peak broadening resulting from longitudinal diffusion is often less than 1%. Deviations from sorption equilibrium can be a serious source of band broadening and arises from diffusion limited processes responsible for the transfer of solute between the stationary and the mobile phases. Band spreading depends on the configuration of the bed material, the velocity of the mobile phase, and particularly the thickness of the stationary phase and is inversely related to the diffusion coefficient of solute in the stationary phase. Narrow bands can best be achieved by coating a stationary phase thinly over a non-porous support (pellicular packing) or using stationary phases of low viscosity hence high rates of solute diffusion. Most recent improvements in column performance have come from the use of bedding materials with maximum interfacial contact between mobile and stationary phases. The relationship between the height equivalent to a theoretical plate, H, and the different physicochemical parameters responsible for maintaining discrete separation of the components of a mixture is illustrated in Fig. 2. It can be seen that the contribution to peak broadening due to molecular diffusion increases markedly as the time of separation increases (decreased flow rate) and mass transfer effects are directly related to flow rate. Since eddy diffusion is relatively unaffected by flow rate the most efficient separation (minimum *H*) is achieved at some optimum flow rate of the mobile phase determined by molecular diffusion and mass transfer effects.

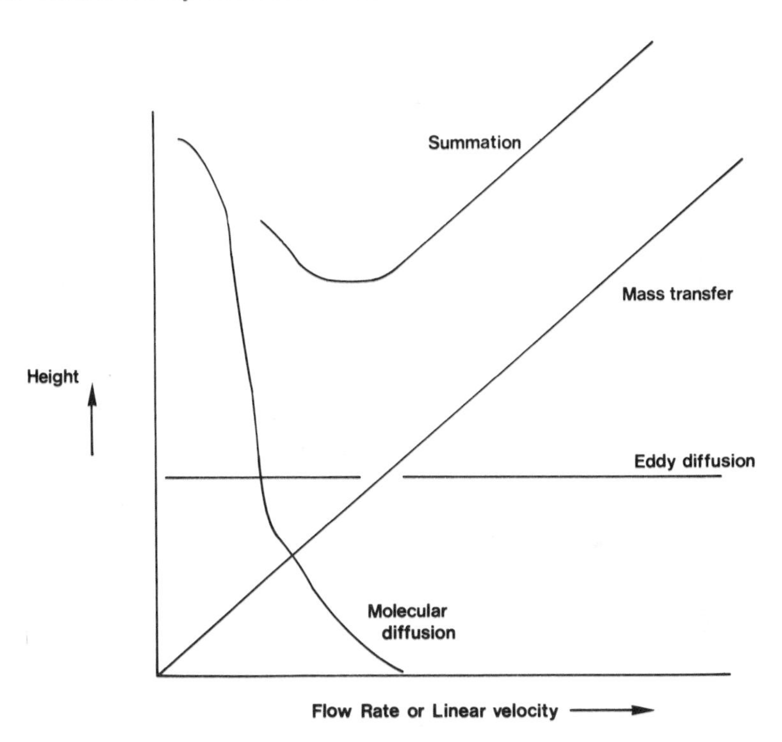

Fig. 2 Relationship between height equivalent to a theoretical plate (*H*) and physicochemical factors responsible for column efficiency as a function of the flow rate of the mobile phase through the column

The profile of peaks can also be affected by the size of the sample applied to the column; this is illustrated in Fig. 3. For sufficiently small samples the peak height is directly proportional to the sample size but retention times remain the same and the separation of the two components is unaffected (chromatograms a and b). When the amount of sample exceeds some critical value a marked decrease in retention time is observed for one or more of the components (chromatogram d). If the sample size is increased still further there is usually a failure to resolve peaks with a further decrease in all retention-time values (chromatogram e). It can be shown that as the concentration and/or volume of sample applied to a column is increased so that the column becomes overloaded the column plate numbers *N* eventually decrease.

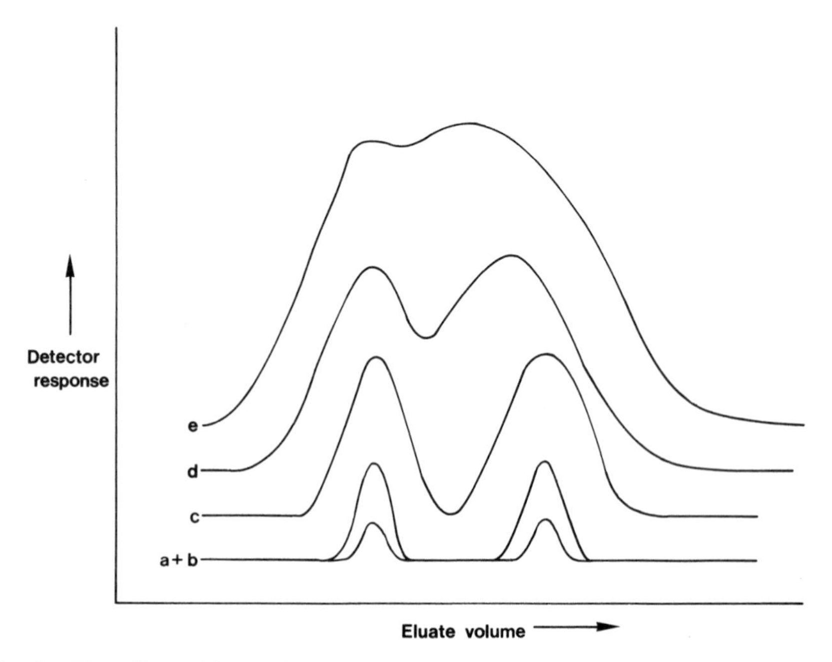

Fig. 3 The effect of increasing sample size on the peak profile and retention times. Sample size increases from a to e

A measure of the chromatographic separation of two peaks is referred to as resolution, R_s. The value of R_s can be calculated from Eq. 4:

$$R_s = 2\,\frac{t_B - t_A}{W_A + W_B} \tag{4}$$

where t_A and t_B are the retention distances and W_A and W_B the peak-base widths of peaks A and B respectively. An example of the measurement of R_s is illustrated in Fig. 4. If the peaks can be regarded as equivalent to Gaussian curves, effectively baseline resolution is obtained for $R_s = 1.5$. The factors contributing to R_s can be seen by substituting for W in terms of N in Eq. 4 giving:

$$R_s = \frac{1}{4}\frac{(\alpha - 1)}{\alpha} \cdot \frac{k_B}{1 + k_B} \;. N \tag{5}$$

Where α is the column selectivity defined as the ratio of the column capacity ratios of substances A and B (k_B/k_A). Each of the three factors contributing to R_s can be optimized, for example, by changing the temperature (affecting diffusion processes) or the nature of the mobile or stationary phases. N can be increased simply by increasing the column length and also by decreasing

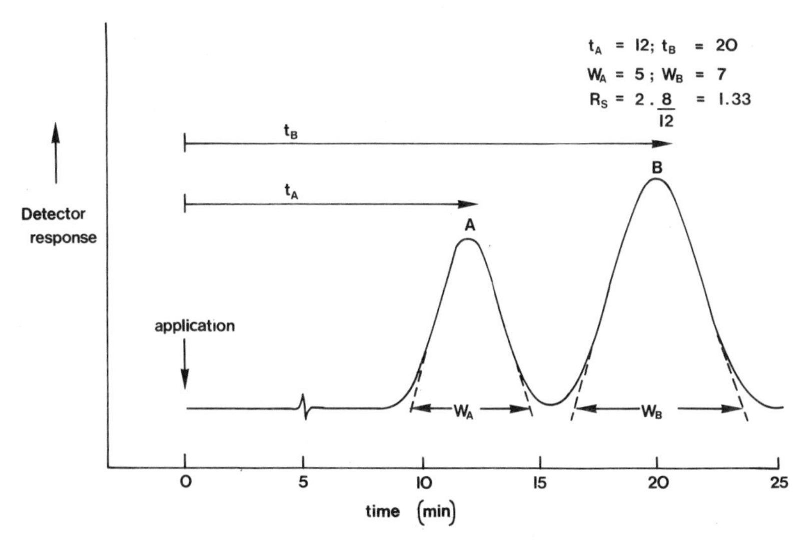

Fig. 4 Determination of the resolution, R_s, of two components, A and B applied to the column as a mixture at time, $t = 0$

particle size or decreasing the flow rate of the mobile phase. Since in practice R_s is a function of time, to achieve the high performance of modern liquid chromatographic systems the necessary flow rates through column beds of small particle diameters is achieved by the use of high solvent pressures applied to the inlet side of column.

4.2.2 Modes of Liquid Chromatography

As can be seen from the flow chart presented in Fig. 5 the ability to keep components of a mixture within well resolved bands relies on various physicochemical factors dealt with in the last section. The ability to separate one component from another, however, depends on their individual chemical interactions with the mobile and stationary phases or, in the case of gel permeation chromatography, simply on molecular size and shape. These interactions can involve a variety of forces and hence depend on the chemical nature of the solutes subject to chromatography and the particular type of solvent and stationary phase employed.

The mode of liquid chromatography appropriate to any particular application depends on the relative sizes of the constituents of the mixture to be resolved and their physicochemical character. These modes and their relevance to large and small constituents are listed in Table 1 and schematically illustrated in Fig. 6. Essentially these fall into four main categories each of which we will examine individually.

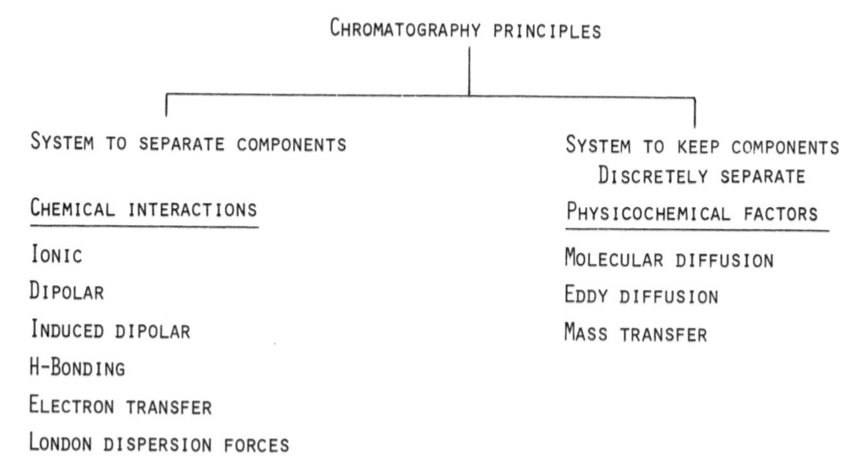

Fig. 5 Principles of chromatographic separation in liquid chromatography systems

4.2.2.1 Gel Permeation or Filtration Chromatography

The principle of this mode is that two compartments are created in the chromatography system one of which is more or less accessible to solute depending on the size. The methods have wide applications to polymers and macromolecular structures including polypeptides, carbohydrates and lipid aggregates. Small constituents which can penetrate into gels which contain a proportion of the solvent but in a stationary mode are retarded in their passage through the column compared to large constituents which are excluded from the gel and are hence confined more to the solvent, representing the moving phase. By monitoring the column effluent an elution profile can be obtained of the different sizes of components of a mixture. The range of molecular sizes that can be resolved by these methods depends on the individual character of the gel; an ideal gel permeation calibration curve is illustrated in Fig. 7 which shows the relationship between log molecular weight as a function of elution

Table 1 Modes of liquid chromatography

Size of constituents	Mode of chromatography
Small components Mol.wt. <1000	Gel filtration Gel permeation Adsorption Partition Ion exchange
Large components Mol.wt. >1000	Gel filtration Gel permeation

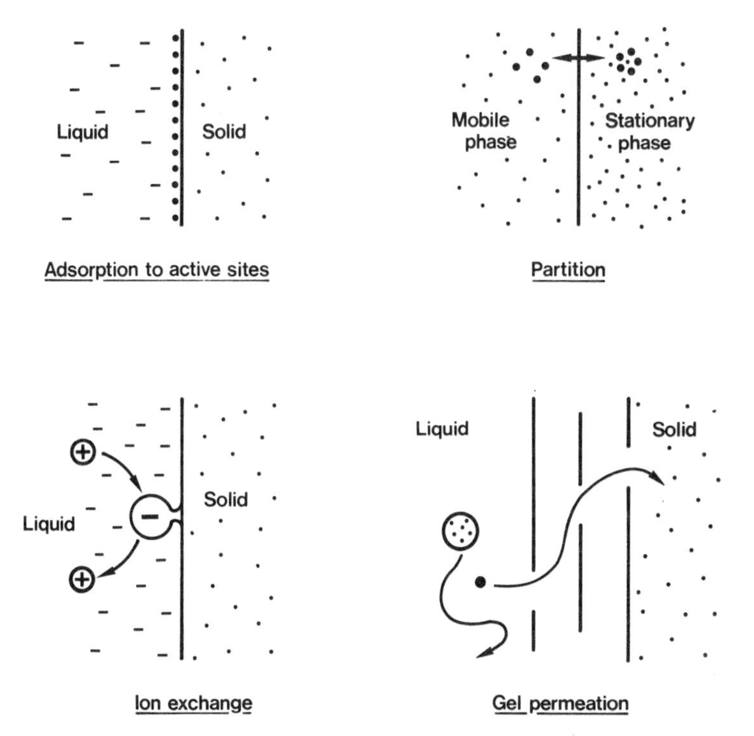

Fig. 6 Schematic illustrations of the principles underlying the various modes of liquid chromatography

volume. The exclusion limit of the gel is where the pores in the gel are too small to allow penetration of solutes above a certain size and the solute is completely excluded from the gel.

A related and chemically non-interactive process in which molecules are separated on the basis of their physical properties is referred to as steric exclusion chromatography. A range of mechanically stable, non-swellable macroporous resins have been developed for steric exclusion chromatography for use with a variety of solvent combinations. Polystyrene divinylbenzyne matrices are examples of such resins. Adequate separation can be achieved by choice of suitable column materials or by using different columns in series to alter the exclusion limits or increase the total pore volume.

4.2.2.2 Adsorption Chromatography

This method sometimes referred to as liquid–solid chromatography relies on differing affinity of components of a mixture for a liquid moving phase and a solid stationary phase. It is often employed for relatively non-polar,

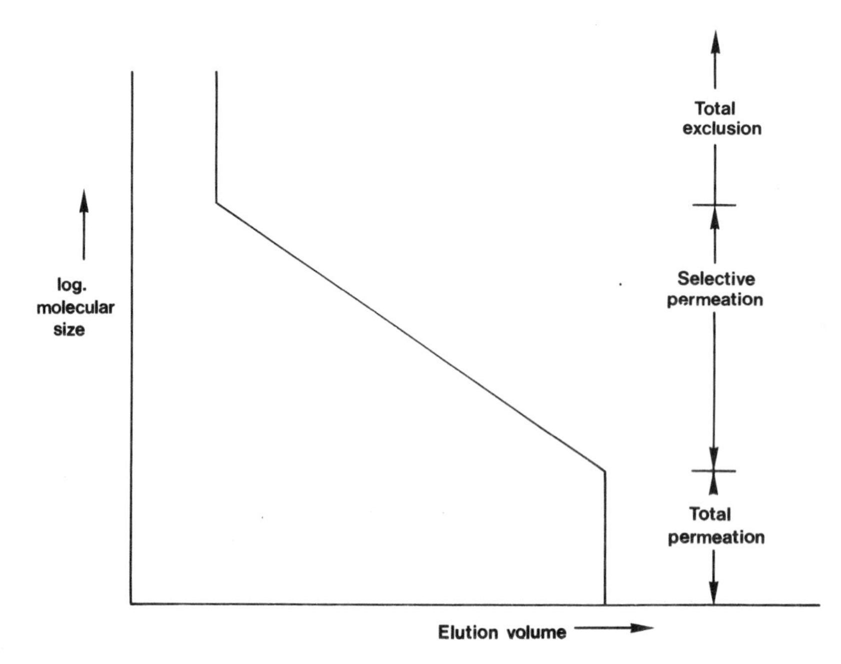

Fig. 7 Relationship between the molecular size and elution volume in an ideal gel column

hydrophobic materials so that the solvent tends to be non-polar while the stationary phase is polar. Silica gel and neutral alumina are commonly employed as adsorbents which can be used with solvents such as pentane and tetrahydrofuran. In general, for a particular material, the more polar the solvent, the shorter the time required for separation.

4.2.2.3 Partition Chromatography

The basis of this system is the partition of solutes between two immiscible liquid phases, one stationary and the other mobile. Immiscibility of the phases may be due to differences in polarity or the covalent bonding of the stationary phase molecules to a solid support. The latter system includes a widely used reversed phase procedure in which a solid support phase such as silica is extensively reacted with mono-, di- or triorganochlorosilanes which cover the polar surface with long C_8–C_{18} hydrocarbon chains. The use of polar solvents such as water–methanol or water–acetonitrile does not remove the covalently bonded hydrocarbon phase from the silica surface which represents an essentially liquid, non-polar stationary phase.

It is possible to separate both ionized and non-ionized solutes using reversed phase chromatography. Suppression of ionic charges by manipulation of the pH of the moving phase, for example, may result in substantial shifts in partition coefficients of solutes between the mobile and stationary phases. More specific interactions between charged groups forms the basis of ion-pair chromatography. This method exploits the principle that the partition of solutes between two immiscible phases can be achieved by formation of ion pairs with suitable counter ions. The general situation can be expressed:

$$A^{+}_{p} + B^{-}_{p} \leftrightarrows AB_{n-p} \qquad (6)$$

where subscript p refers to be mobile polar phase and n–p to the non-polar stationary phase. The equilibrium position is determined by the nature of the individual ions and the extent to which the ion pair partitions into the non-polar stationary phase.

4.2.2.4 Ion Exchange Chromatography

This mode of separation is applicable to charged solutes which can be separated on the basis of their strength of binding to oppositely charged groups presented on the stationary phase. The principle of ion exchange chromatography can be exploited in combination with gel permeation or exclusion chromatography to resolve solutes differing by a combination of charge and steric characteristics. The methods are generally applicable to aqueous systems but organic modifiers such as ethanol or methanol can, in certain cases, be added to the aqueous mobile phase to improve the efficiency of separation of particular solutes. The presence of other ions also alters the affinity of charged solutes to fixed charges on the stationary phase. More efficient separations are usually achieved when the ionic strength of the stationary phase is low. The rate of exchange of solutes between the mobile and stationary phases can also be affected by temperature. The pH of the system can be manipulated to alter the charge of the solutes depending on the particular pK_a but charges on the stationary phase are also modified by the solvent pH.

Irrespective of the mode of liquid chromatography employed some general methods can be applied to increase the resolution of closely related solutes in a mixture. One often successful method is to form covalent derivatives of the solute as is commonly performed to obtain volatile compounds for separation by gas chromatography. Derivatization can increase the stability of solutes under conditions appropriate for liquid chromatography and improve their separation characteristics in a mixture. Another common application of derivatization is to form compounds that can be detected in trace concentrations hence improving the sensitivity of the method.

With many modes of chromatography the performance can be significantly improved by altering the characteristics of the mobile phase during the chromatographic process. Normally the composition of the mobile phase entering the column is held constant and this procedure is referred to as isocratic elution chromatography. Additional resolving power can often be achieved by a gradient elution process in which the composition of the mobile phase is programmed to change with time. One advantage of this procedure is that it overcomes, to some extent, the general elution problem associated with mixtures of compounds having widely differing retention times. Under isocratic conditions, for example, the first eluted material is often poorly resolved whereas the last eluted bands tend to be considerably broadened, the solutes excessively diluted and difficult to detect accurately. This can be overcome by programming the mobile phase. Gradients may be programmed for monotonous changes in the mobile phase or for discontinuous changes that involve a series of isocratic gradient steps.

4.3 HPLC Equipment and its Operation

4.3.1 The HPLC System

The components of an HPLC system may be divided into five separate items comprising: the mobile phase delivery system (pump), the injector, the column, the detector and finally, the recording system. The overall arrangement of these components in a typical system is shown as a block diagram in Fig. 8.

The design trend of recent years amongst the major manufacturers of HPLC equipment has generally been to avoid the totally integrated system where all

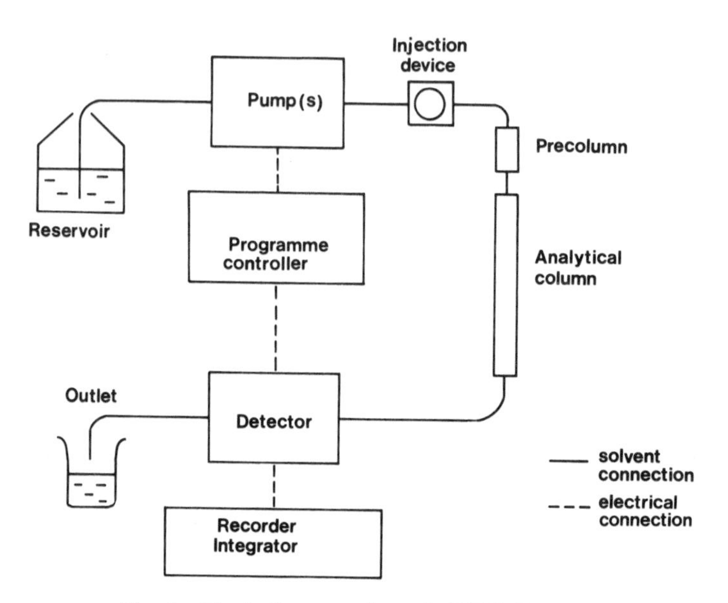

Fig. 8 Block diagram of a typical HPLC system

the components are built into a basic shell and instead a modular concept has been adopted where individual, compatible units are linked together. Whilst, the latter option results in a physically larger system, the advantages are that it provides easier access for fault diagnosis and allows the individual components to be changed as technological advances are made in respect of the individual components.

In the following sections relating to the individual components of HPLC equipment, various manufacturers have been mentioned, whilst others are not. This is not meant to indicate any preference for particular instrumentation but simply that equipment with which the authors are familiar has been included.

4.3.2 Pumps and Gradient Systems

The basic requirements of an HPLC pump are to produce a constant, reproducible flow of solvent through the column (usually between 0.1 and 10 ml/min) at high pressure (up to 5000 psi/35 $N.m^{-2}$). The two pumping modes which are employed are constant volume flow and constant pressure flow. Constant pressure pumps have the considerable advantage of being pulseless but the flow rate is greatly affected by changes in column resistance or variation in solvent viscosity. Two types of constant pressure pumps are available. The simpler system uses gas cylinder pressure to drive solvent out of a metal coil but this system has the disadvantage of difficulty in changing solvents. The second type of constant pressure pump most widely used is the gas-driven pressure intensifier. This pump uses gas at relatively low pressure acting on a piston of large surface area which is directly connected to a hydraulic piston of small surface area. The higher the ratio of the two surface areas the greater is the pressure amplification. The gas-driven pressure-intensification pumps have found extensive use in preparative HPLC.

Constant flow pumps are also of two types. The most common are mechanical reciprocating devices and usually they have an electrically driven piston working directly on the solvent. The major disadvantage of these pumps is that they produce a pulsatile flow which can adversely affect flow-sensitive systems. This disadvantage may be overcome in several ways; first, the pulses may be damped by using either a tube with an air space, or a flexible bellows, so that during the pulse phase some of the energy is taken up by the damping device, while during the refill phase of the pump the energy is released, thus helping to maintain the pressure in the system. The second method of overcoming pulsatile flow is the use of dual- or triple-headed pumps with the pumping heads arranged out of phase so that one head is pumping whilst the other head is filling. This considerably smooths the solvent flow. The third and more recent method uses electronic pulse compensation circuitry. In this type of pumping system the electronics automatically speed up the pump motor during both the refill and the piston chamber pressurization periods, resulting in a great reduction in the period of time during which the piston is not delivering solvent to the outlet. The resulting flow rate is essentially pulse free.

The second and less commonly employed type of constant-flow pump is the motor-driven syringe which whilst possessing the capability of maintaining very accurate, pulseless flow rates at high pressures, have a finite volume and therefore have to be refilled. Also they are relatively expensive.

In summary, piston pumps are the most commonly and conveniently employed due to their low price, high reliability and ease of use with gradient elution systems.

While it is often necessary to use only a single solvent to separate a mixture, i.e. isocratic separation, more complex mixtures may require a gradient elution where the eluting power of the solvent is gradually increased as the chromatography proceeds. Gradient systems are subdivided into high and low pressure systems. In low pressure systems the gradient is generated before the solvent enters the pump and requires that the pump has a small dead volume where a rapid change of solvent is possible; for these reasons, low pressure systems are generally used with reciprocating pumps. In low pressure systems gradients may be generated using relatively simple, cheap designs which alter the flow from different solvent reservoirs prior to a mixing chamber before the pump (see e.g. Perrett, 1976).

High pressure gradient systems usually employ two high pressure pumps whose output of solvents from two separate reservoirs is controlled. The flow rates from each of the pumps is regulated by an electronic programmer thereby allowing gradients of different profile to be produced. Recent advances in microprocessor technology have allowed increasingly sophisticated controllers to be integrated into solvent delivery systems with for example the Altex 421 CRT Controller being capable of storing and executing more than 99 different programmes. While these high pressure systems increase the flexibility and potential of an HPLC it should be realised that because they require two pumps in addition to a programme controller, they are relatively expensive.

4.3.3 Injectors

The choice of injection systems is between syringe and valve devices. Syringe injection has the advantage of delivering the sample at the top of the column without any dilution, thereby increasing the potential for sample resolution. Syringe systems in practice, however, pose three major problems. First, they inject the sample in-line and are therefore exposing the relatively fragile syringe to high pressures. Second, the sample is injected through a septum usually made of silicone rubber; material from these septa can both block the injection syringe and contaminate the top of the column. A third and less commonly encountered situation is that some organic solvents, which are finding increasing usage in reverse phase HPLC, may attack the septum material.

The second type of injection system utilizes a valve which usually contains a loop into which the sample is injected and which can be isolated from the rest of the HPLC system (see Fig. 9). After the sample has been injected the valve is mechanically actuated to bring the loop (and sample) in-line with the HPLC system, thereby resulting in the sample being flushed onto the top of the column. Injection valves are usually recommended for use with commercial HPLC systems and provide the major advantages of precision, reliability and ease of operation. Furthermore, recent advances in valve technology which have resulted in a reduction in the dead volume between injection and the column, allow column efficiencies to approach those found with syringe injection.

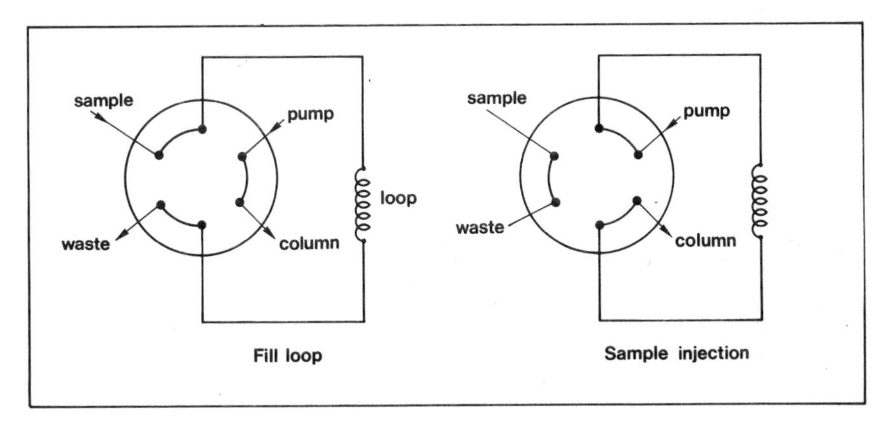

Fig. 9 A diagram of a loop injection system for the application of precise volumes of sample to the column

An increasingly important advantage of valve injection systems is their ease of use with automatic injection systems which use a pneumatic actuator to switch the loop from its isolated to in-line positions. The automatic injection systems currently available are of two basic types: the first, uses a turntable sampler (e.g. Waters Associates, Applied Chromatography Systems Ltd, Kontron Instruments Ltd) and has the advantage of handling a relatively large number of samples. The second, simpler system uses a gravity feed for the samples (e.g. Altex) and is considerably cheaper than the turntable devices. The obvious advantage of these automated delivery systems is that they provide increased cost-effective use of expensive equipment by allowing its overnight use for routine analyses and daytime operation for research and methods development.

4.3.4 Columns and Column Packings

The most important single item of equipment in an HPLC system is the column which contains the packing material, as even the most technically advanced systems will be unsuitable without efficient sample resolution. The relatively high pressures currently employed in most HPLC systems have led to the use of columns constructed of a good grade of stainless steel instead of the glass columns previously used which had a maximum pressure tolerance of ca. 800 psi. The metal making up the column should have a fine, high quality finish as surface irregularities can result in both a decreased performance and non-reproducible behaviour from column to column due to differences in the packing efficiency.

Most analytical columns are 10–20 cm in length and 4–6 mm internal diameter and are packed with 5–10 nm particles. Longer columns require corresponding increases in internal diameter to avoid the infinite diameter effect (Knox, 1976) where the sample travels down the wall of the column. It has been suggested that there is little difference in the efficiency of resolution between irregular and spherical particles, but it should be realized that the latter will allow columns of greater permeability with a corresponding decrease in the pressure requirement for a given flow rate. Most packings currently used are based on silica due to its versatility as an adsorbent. The surface of the silica is modified by chemically bonding on different modifying groups. Ideally the particle covering should be a monolayer and there should be no unreacted silanol groups exposed to the solvent as these may cause tailed peaks when polar compounds are being resolved.

A large number of different types of modifier bonding are now available including octadecanoylsilyl (ODS) groups, short alkyl silyl (SAS) groups, aminopropylsilyl (APS) groups, amine (NH_2) groups and diol groups. These various bonding materials are used in the different modes of chromatography described earlier. More recently considerable advances have been made in the packing materials used in steric exclusion chromatography. These materials are now supplied in different exclusion ranges depending on the molecular weights of the materials to be resolved. These packings consist of microporous resins which ideally should be rigid and resistant to swelling and shrinkage in a wide variety of solvents.

Many of the prepacked columns are relatively expensive (ca. £150 for a reverse phase column containing 15,000 plates), and it is therefore advisable to introduce a small precolumn (ca. 40 mm × 3 mm i.d.) between the injector and the analytical column. The precolumn is packed with microparticulate materials similar to those in the analytical column. Precolumns are particularly recommended for applications where compounds may accumulate on the top of a column, e.g. samples containing high molecular weight or polar compounds. Provided there is a minimal dead space between the injector, the precolumn, and the analytical column, there should be little loss of resolution and there may even be an increase in the total plate count for the system.

Two recent advances in column technology are worth mentioning to the potential HPLC user. The first concerns a recognition of the importance of temperature control to the efficiency of a column and which has led to the development of thermostatically controlled column heaters (e.g. Bioanalytical Systems). It has been demonstrated that a change in column temperature as small as 2 °C can significantly alter the behaviour of a column and therefore, in environments where the ambient temperature may fluctuate a column heater is essential for reproducible results. Furthermore, an increase in column temperature can result in a significant reduction in the capacity factor of a column and may thereby decrease the consumption of organic solvents.

The second recent advancement is the use of radial compression systems (Waters Associates) which are reported to improve the speed of separations whilst maintaining resolution. To achieve rapid high efficiency separations a short column is required together with an extremely narrow diameter column packing. Unfortunately, at the flow rates required for rapid separations the back pressure exerted by such a column would be unacceptably high. Radial compression overcomes these problems by the use of a flexible-wall column held under uniform hydraulic pressure thereby radially compressing the packed wall structure. This causes the outer wall of the column to conform to the outer contours of the particle bed, resulting in a short, high efficiency column with comparatively low back pressures. While the initial outlay for the compression device is relatively high (approx. £800), the columns are considerably cheaper than their steel equivalents.

4.3.5 Detectors

The detector modules of an HPLC system monitor the column eluent. They may be broadly divided into two categories:

(1) solute property detectors which determine a specific physical property of the eluting solute which is not present in the mobile phase; and
(2) bulk property detectors which determine the change in a physical property of the mobile phase.

Clearly the detector to be employed is dependent upon the nature of the solutes to be detected, however, it may be generally stated that an appropriate and good detector should have a high signal response coupled with a low background noise level. In any detection system the drift characteristics, which may have a multitude of causes, must be quantified and where possible minimized. It should also be recognized that the volume of the flow cell within the detector should be low (usually 5–10 μl) to preclude any disturbance to the eluting peaks.

At present the most widely used solute property detector measures ultraviolet (UV) absorption. Originally these detectors were based upon the emission characteristics of a mercury source, but more recently variable wavelength detectors have become increasingly popular. As UV detectors measure a solute property they are relatively insensitive to both flow rates and temperature variations; however, these detectors are subject to pressure effects and so when a reciprocating pump is being used a pulse damper should also be employed. The sensitivity of variable wavelength UV detectors has increased steadily over recent years due to improvements in both the stability of the lamps and also in the electronics. The minimum limit for detection with UV spectrophotometers is obviously dependent upon the extinction coefficient of the solute but is of the order of 10^{-11} M. Detection and characterization of an unknown sample may be facilitated by the use of variable wavelength double beam scanning spectrophotometers which can trap an eluted peak within the flow cell and then scan its absorption characteristics. Similarly, in the cheaper single beam instruments, the same sample can be run twice, each time at a different wavelength to facilitate characterization.

A second solute property detector which is popular is the fluorimeter. These instruments measure the fluorescent energy at a specific emission wavelength of a solute which has been excited by radiation at a different wavelength. While fluorimetric detection is very sensitive (often subpicomolar minimum sensitivity), the system is very specific and not all compounds are fluorescent. Often this problem can be overcome by coupling a non-fluorescent solute with a fluorescent compound. This process of derivatization can be carried out either before resolution of the compounds on the column (precolumn derivatization) or after resolution on the column (post-column derivatization).

A recent advance in fluorimetry which has been applied to HPLC is the use of a multichannel detector for the acquisition of two-dimensional fluorescence data (Warner *et al.*, 1975). In this technique fluorescence intensity is monitored as a function of multiple emitting and exciting wavelengths to produce a 200 nm by 200 nm fluorescence matrix. Subsequently, both qualitative and quantitative analyses can be applied to these fluorescent scans. These two-dimensional fluorescence detectors are not yet commercially available.

The most widely used bulk property detector is the refractive index (RI) detector. These instruments continuously monitor the difference in RI between the column eluent and a pure mobile phase. The presence of sample constituents in the mobile phase causes the RI to change thereby making detection possible. The limitations of these RI detectors are their lack of sensitivity and susceptibility to temperature change. The minimum sensitivity of these detectors is ca. 1 nmole; however, to achieve such sensitivity the temperature control must be of the order of $\pm 10^{-3}$ °C. To maintain such temperature stability a heat exchanger is usually incorporated but this increases the volume of the detection system and so may result in a broadening of the eluted peaks. One further limitation of these detectors is that they cannot be used with a gradient elution system. RI detectors are therefore of restricted use for analytical purposes where a low detection limit is required but may be usefully employed in either preparative or semi-analytical HPLC.

The most universally applicable bulk detection system is the transport detector. This instrument requires the transport of the column eluent by a moving wire to a flame ionization detector of the type used in GLC. The system may be used to detect either lipids, proteins or carbohydrates. The minimum level of sensitivity is of the same order as for RI detectors. One of the critical problems with this technique is its susceptibility to flow rate.

The detection system which has shown the most rapid advances over recent years is the electrochemical or amperometric detector. In this system, the resolved components from the column pass through an electrochemical cell where they are subjected to oxidation or reduction at the surface of an electrode at a selected and constant polarization voltage. The very small current produced by this electrochemical reaction is measured and amplified to produce the eluting peaks. The current produced is proportional to the amount of material oxidized/reduced. The detection system is extremely sensitive showing a minimum sensitivity in the subpicomolar range.

There are three types of electrode materials which are usually used in electrochemical detectors and these include carbon paste, glassy carbon, and a mercury/gold amalgam. The choice of electrode varies with the solutes to be resolved and the solvents being employed but to optimize detection conditions the electrode should:

(1) provide a low background current at the operating potential;
(2) possess a wide range of operating potentials; and
(3) give a rapid electrochemical reaction with the solutes being detected.

A further factor to be considered is that while carbon paste electrodes are often optimal for the aforementioned criteria they are susceptible to direct chemical reaction with certain organic solvents such as acetonitrile. Thus, the choice of electrode is a balance between several individual factors.

A problem which can exist in electrochemical detection is where a mixture contains several substances with very different redox properties all of which are present at low concentrations. It would be impossible to concurrently determine them all as a relatively low but specific electrode potential would be required to reduce the background current. To overcome this problem it is possible to form a dual electrode where two adjacent working electrodes are present within the same thin-layer detector cell each operating at a different potential.

The detection of radioactive material is an additional means of extending analysis and of providing increased sensitivities. A large number of radiolabelled compounds are available at high specific activities to facilitate their use in biochemical and pharmacological studies. While few commercial in-flow detectors are presently available for use in HPLC systems, it is possible to collect the eluent from the column in small volumes with the aid of a fraction collector. Scintillation fluid may then be directly added to the fractions prior to measurement of the radioactivity present by using a scintillation counter.

4.3.6 Additional Equipment

The most important additional item of equipment is a recorder to provide a quantitative record of the elution. The cheapest method of obtaining a record is to connect a simple variable-input chart recorder to the detector output, thereby allowing the sensitivity of the recording to be adjusted. Quantitation of the compounds in the elution profile is achieved by measuring either peak height or peak area. The latter is more frequently used in gradient elutions where the baseline may change.

In experiments where a large number of samples are being processed, or where automatic sampling is in operation, it is worthwhile to consider the use of an integrator which generates numbers from data points related to the chromatogram. These numbers are subsequently used for quantitation. It is important that the integrator being used should be able to correct for both baseline variations and background noise. Once the integrator is programmed to achieve its function it not only removes the laborious routine of simple, repetitive calculations but, as with automatic sampling devices, it makes the HPLC systems labour intensive.

The use of a fraction collector has been described above for the automatic sampling of the column eluent. Such collectors can remove the tedium from fraction collecting and they are invaluable for use in preparative HPLC.

A common problem arising in HPLC analysis is that of solvents degassing within detector flow cells which results in the formation of bubbles and a ruined chromatogram. Degassing occurs where relatively low pressures (below 2000 psi) are being employed and may be avoided by installing a degassing device into the HPLC system. Several devices are commercially available.

Finally, it should be emphasized that no chromatographic separation will be satisfactory without pure solvents. If impurities are present they may cause high baseline drift, excessive background noise and may even make analysis impossible. Fortunately, a number of chemical suppliers are now marketing HPLC grade solvents which are ideal for analysis even at low sensitivities.

4.3.7 Column Packing Techniques

It is often advisable to consider packing columns oneself as commercially packed columns are usually £150–200, while after the purchase of the initial equipment (approx £200) it is possible to pack columns for less than £30.

The two principle techniques of column packing are dry packing and wet packing. Dry packing is generally only used with particles >20 nm in diameter and the most commonly employed technique for packing is the tap-and-fill method. The development and general use of microparticulate packings has resulted in a trend towards the use of wet packing or slurry packing techniques of which a number exist. A brief description of one of these will now follow.

The importance of the surface finish of the column has already been described (Section 4.3.4) but it should also be noted that the column should be extensively degreased by using several organic solvents. The cleaned column should be dried and either a stainless steel or PTFE frit fitted to the end of the column to retain the packing. The pores of the frit should be less than half the size of the particles making up the column packing; however, since most commercial packing materials contain particles which are smaller than the average particle size it is necessary to remove them to prevent clogging of the frit. To achieve their removal, the packing is suspended in a suitable solvent and allowed to sediment under gravity for about 15 minutes. The supernatant containing the finer particles is carefully removed and the process repeated until the supernatant is clear. The sedimented particles are then dried and are ready for use.

Essentially all slurry packing techniques require the same equipment and ideally use a constant pressure pump capable of operating up to about 5000 psi in series with a reservoir to hold the slurry of suspended packing, which is connected to a short precolumn in front of the main column to be packed. To pack a 10 cm ×5 mm internal diameter column with 5 nm diameter particles a packing pressure of approx. 3000 psi is recommended. The size of the slurry reservoir is clearly dependent on the volume of the column to be packed but for the column described a 25 cm^3 container should be adequate. The precolumn should be 2–3 cm long and should have the same internal diameter as the main column (i.e. 5 mm). The precolumn is necessary to maintain an equal particle packing density within the main column and there should be no dead space between the two columns.

The packing material should be suspended in a mixture of suitable organic solvents with which the bonded surface material of the packing will not react. Organic mixtures which are frequently used include high-density halogenated hydrocarbons or methanol, plus aqueous sodium acetate. The particulate suspension should be ultrasonicated for at least 5 min. The column and precolumn are filled with the same organic mixture as that in the particle suspension and then the slurry is tipped into the reservoir. The fittings are now connected up to the pump and a different organic solvent is pumped through the reservoir and column at about 3000 psi until approx. 100 ml of the second solvent have passed through. The pressure in the system is allowed to return slowly to zero (for example, by isolation of the pump from the system) and then the column is connected directly to the pump and flushed with 400 ml hexane at low pressure (e.g. 100 psi). Once again the pressure is allowed to drop and then the column may be removed. The packed column is then sealed and protected with a stainless steel frit which is positioned on top of the column packing. Where injection directly onto the column is to be used in preference to an injection valve, it is also necessary to add about 5 mm of glass beads (100 μm diameter) above the frit. The column is then sealed with a PTFE plug and is ready for testing.

It is essential to assess the performance of any column after either packing or purchase. This will provide an indication of both efficiency of the column and also the rate of deterioration of the column. As described earlier (Section 4.2.1) the most meaningful criterion of column performance is the plate number and this should be assessed routinely for any column.

4.4 Applications of HPLC

The increase in popularity of HPLC may to a large extent be attributed to the expansion in the number of potential applications to which it may be put. The improved sensitivity of detectors and the advent of commercially produced electrochemical detectors have contributed significantly to the increased capabilities of HPLC.

Before discussing specific examples of the different applications of HPLC it would be appropriate to consider the nature of the sample to be injected. The major criteria to be observed are that:

(1) the sample is at a sufficiently high concentration to allow its detection in the volume injected;
(2) the sample is free of contaminating material which would damage or block the column;
(3) the sample is suspended in a similar solvent to that used for the column eluent; and
(4) a suitable internal standard is present to allow quantitation of the sample peaks.

The first two of these criteria may often be met by organic extraction, followed by concentration through evaporation, whilst the internal standard may be decided upon by experimental trial. Specific examples of some of the biochemical, toxicological and clinical applications to which HPLC has been put over recent years may now be considered.

4.4.1 Drugs

The expansion of HPLC applications applied to drug analysis has probably expanded faster than any other major analytical area (Pippenger, 1979). A large number of clinically employed drugs show a significant correlation between blood concentrations and therapeutic effectiveness; thus, drug monitoring can aid the clinician in understanding why a patient is not responding to a particular drug régime or why certain undesirable side-effects are apparent. Such considerations are clearly of great importance where a raised level of a drug may be toxic. An advantage of HPLC over other techniques such as radioimmunoassay in the therapeutic monitoring of drug concentrations is its relative cheapness. Examples of some drugs which are in common clinical usage, where the blood concentrations are of major importance include tricyclic antidepressants, theophylline and its derivatives, β-adrenergic blocking agents and anticonvulsants. Tricyclic antidepressants such as imipramine and amitriptyline are the most widely used drugs for the treatment of mental depression and may be readily separated within 10 minutes on a suitable reverse phase column and monitored at 254 nm with a suitable UV detector. Theophylline is an alkaloid of the xanthine group and is used for symptomatic relief of acute bronchial asthma and for bronchospasm associated with chronic bronchitis and emphysema. Resolution of theophylline on a reverse phase column is necessary because a number of other drugs and metabolites absorb at the optimal detection wavelength of 214 nm. β-Adrenergic blocking agents such as propranolol are widely used in the treatment of cardiac arrhythmias, angina pectoris and hypertension, but it is believed that at elevated plasma concentrations propranolol may be cardiotoxic. Propranolol and its metabolites may be resolved on a reverse phase column and detected by its fluorescence emission at 216 nm. Valproic acid is one of the most commonly employed anticonvulsants and is particularly useful in paediatric medicine. Valproic acid is readily separated from other metabolites within 10 minutes on a suitable reverse phase column and can be detected at 246 nm with a UV detector.

4.4.2 Steroids

A wide range of reverse phase columns are capable of providing excellent resolution of most biologically active steroids; however, with certain adrenocorticotrophic hormones it is necessary to carefully select the correct packing and mobile phase. Clinical monitoring of steroids is of considerable importance notably in pregnancy where knowledge of the urinary oestriol levels are of considerable use in the management of high risk patients facing complications such as fetal growth retardation, rhesus immunization or diabetes. Normally, oestriol levels rise during pregnancy and a decline is indicative of placental malfunction. Large differences in basal and pregnant values for plasma and urine levels occur between individuals. It is therefore necessary to perform several tests on a given individual. Previous methods of detection of oestriol have relied upon HPLC with UV or fluorescence detection, and also radioimmunoassay (RIA). The first two detection systems are relatively insensitive while RIA suffers from a lack of specificity. These problems in detection have been superseded by two alternative detection techniques, which may also be used for the detection of other steroids such as oestrone, oestradiol and several of their metabolic derivatives. The first of these techniques employs electrochemical detection which has been described earlier (Section 4.3.5) and relies upon the oxidation of the phenolic ring of the steroid at a carbon paste electrode; the sensitivity is sufficient to detect the oestriol in as little as 1 ml of a urine sample. The second technique which has been reported uses the separation capability of liquid chromatography combined with the sensitivity of RIA. Thus the chromatography reduces the confusion due to cross-reactivity and RIA provides as extremely sensitive detection technique for oestrogens (Sivorinovsky, 1980). Initially, to establish the retention times for the individual oestrogens it is convenient to use commercially purchased samples which may be monitored by UV detection.

4.4.3 Biogenic Amines

The biogenic amines and more specifically, the catecholamines, provide a measurable index of sympathetic nervous system activity through measurement of their plasma or urinary concentrations. It is frequently desirable to measure the effects of certain drugs on the sympathetic nervous system but while urinary catecholamines will provide an indication of the overall physiological changes which a drug may incur, it is the plasma concentrations which change more rapidly and provide a more accurate index of the organism's response to a stimulus.

Until recently the most popular modes of measurement consisted of either fluorescent detection, gas chromatography together with either electron capture or mass spectrometric detection or radioenzymatic assay. Each of these techniques is either extremely time consuming, expensive and/or non-specific. More recently, HPLC in combination with electrochemical detection has superseded all the aforementioned methods and probably no other application of electrochemical detection has found more widespread use than the determination of the various biogenic amines and their metabolites in nervous system tissue and physiological fluids (Freed and Asmus, 1979).

A suitable reverse phase column together with appropriate ion-pairing agents in the mobile phase can elicit a separation of noradrenaline, adrenaline and dopamine within 20 minutes, and a minimum detection limit of 20 pg has been reported. Similarly, 5-hydroxytryptophan (serotonin) and its major catabolite 5-hydroxyindole-3-acetic acid are readily separable on an appropriate reverse phase column and again the lower detection limit is ca. 20 pg.

4.4.4 Nucleotides, Nucleosides and Bases

Nucleotides are negatively charged over a wide pH range and so anion exchange has in general been the chosen method of separation together with UV detection at 254 nm (Brown *et al.*, 1979). However, the charge differences between the mono-, di-, and triphosphates have meant that a gradient is necessary with a relatively high ionic strength being required to elute the triphosphates. Most pump manufacturers recommend that high concentrations of halide salts should be avoided to prevent corrosion of the stainless steel parts of the pump and consequently phosphate salts have been used. It should be noted that until recently quantitation of less than 1 nanomole quantities of nucleotides in a gradient system was extremely difficult because of a high baseline rise in the detection systems due to impurities in the phosphate salts. This problem has been resolved by the manufacture of an extremely pure phosphate salt (Aristar, British Drug Houses).

A second approach to the problem has been to use a reverse phase column and to neutralize the negative charge on the nucleotides by including an ion-pairing agent such as the tetrabutylammonium ion in the mobile phase. Under these conditions a gradient is still required but the concentration of potassium phosphate can be reduced from ca. 0.6 M to 0.02 M thus providing a considerable financial saving.

The resolution and detection of nucleosides and bases may be achieved using similar procedures to those described for the nucleotides with the modification that the concentration of potassium phosphate should be further reduced and the pure methanol which is used in the final phase of the nucleotide separation is replaced by a methanol–water mixture.

4.4.5 Lipids

Until recently HPLC analysis has not been widely employed in lipid analysis mainly because most lipid molecules do not absorb in the ultraviolet. The increased sophistication of commercially available refractive index detectors, however, has provided a useful tool for lipid analysis and may be applied not only to fatty acids and their derivatives but also to larger molecules such as triglycerides. It has been demonstrated that fatty acid esters, both saturated and unsaturated, may be readily separated using reverse phase columns and even the *cis* and *trans* isomers of unsaturated fatty acid esters may be resolved (Carr, 1974).

The limited sensitivity of refractive index detectors will continue to pose a problem for quantitation of low nanomole amounts of lipid but this may be overcome by precolumn derivatization with compounds which absorb strongly in the UV. This technique has been successfully applied to the analysis of long chain fatty acids where they were converted to 2-naphthacyl esters or phenacyl esters. The sensitivity of detection using this technique was increased to less than one nanomole of fatty acid.

4.4.6 Carbohydrates

Resolution of carbohydrates such as glucose, galactose, mannitol and sorbitol can be achieved on a simple cation exchange column. Selectivity of the resolution is obtained by optimizing the ionic form and cross-linkage of the resin. Water may be used as the eluent and extraction of the samples is by simple water dilution or extraction. There have also been recent reports of the resolution of oligosaccharides and polysaccharides using size exclusion chromatography. The eluent used in these systems was distilled water containing 0.05 M strong electrolyte. Detection, as in lipid analyses, is by refractive index changes and therefore lacks sensitivity. Similar elution conditions may be applied to glycoprotein carbohydrates such as *N*-acetylglucosamine and *N*-acetylgalactosamine. HPLC will thus provide a satisfactory resolution of carbohydrate moieties but is suitable only where sensitivity of detection is not a problem.

4.4.7 Amino Acids, Peptides and Proteins

There has been an increasing trend in biochemistry in recent years towards the determination of the amino acid sequence of very small amounts of protein. The most popular sequencing method has been the Edman degradation whereby amino acids at the N-terminus of the protein are successfully reacted with phenylisothiocyanate to form phenylthiohydantoin (PTH) amino acid derivatives. The PTH amino acids are easily resolved on a reverse phase column and can be detected by their UV absorption at 254 nm. A gradient elution is used with the second solvent (acetonitrile plus phosphoric acid) containing an organic modifier (tetrahydrofuran) to shorten the retention times while allowing sufficient water content in the mobile phase to take advantage of hydrophobic selectivity. Under these conditions the 20 major amino acids may be separated within 40 minutes (Savage, 1979).

The separation of water-soluble peptides and proteins can be achieved using size exclusion chromatography with the larger molecular weight proteins having the shortest retention time. The eluent employed is usually a low ionic strength aqueous buffer in combination with UV detection at 280 nm. It should be noted that since the molecular conformation of proteins is affected by changes in pH, retention times in size exclusion chromatography will also be affected; thus, the pH of the eluting buffer should be rigidly maintained.

4.4.8 Environmental Toxins

Recently, information on the biological and environmental properties of a number of pesticides, herbicides, fertilizers and naturally occurring toxins has led to an increased interest in the study of both their metabolism and toxicology. Consequently it has become necessary to accurately quantitate these potential biohazards which often only occur in very low concentrations. A few of the applications in environmental studies where HPLC has been usefully used will now be considered.

A number of aromatic amines (e.g. aniline, benzidine, *o*-toluidine) have been reported as potential carcinogens, and many pesticides yield various aniline residues as breakdown products. It is therefore important to evaluate the identity and concentration of these aromatic amines and their derivatives and quantitation is conveniently achieved by separation of the compounds on a reverse phase column followed by electrochemical detection using a carbon paste electrode (see Riggin and Howard, 1979).

An alternative detection system has been used for the detection and characterization of polychlorinated azobenzenes. They can be produced either during microbial degradation of commercial chloroanilide herbicides or during the industrial synthesis of these herbicides and are therefore potentially both industrial and environmental toxins. Resolution of the polychlorinated azobenzenes is on reverse phase columns and they may be detected by their UV absorption at 254 nm.

Fluorimetric detection after post-column derivatization with *o*-phthaldehyde has been used to identify *N*-methylcarbamate insecticides. This method allows the quantitation of insecticides such as Methomyl and Aldicarb to a lower limit of approximately 500 pg per injection which corresponds to a concentration of about 30 ppb which is well below the limit of sensitivity normally required.

4.5 Conclusions

High performance liquid chromatography, as we have seen, has wide application in the separation of molecules of biological interest. New advances in systems for detecting trace amounts of various solutes have widened and will continue to widen the application of the method.

One of the obvious examples of the advantage of HPLC over its main rival, gas chromatography, is that the mobile phase is involved in the equilibrium distribution of the molecules subject to chromatographic separation. This means that apart from the primary equilibrium established between the mobile and stationary phases, secondary chemical equilibria take place between solute molecules and components present in either mobile or stationary phase that can be exploited by the selection of appropriate chromatographic conditions.

Marked changes in the retention characteristics of particular compounds can be achieved, for example, by changes in the ionization state, conformation or the reversible formation of complexes with altered partition coefficients. Continued exploitation of these advantages will doubtless continue to contribute to the versatility of the HPLC technique.

References

Brown, P. R., Hartwick, R. A., and Krstulovic, A. M. (1979) Analysis of blood nucleotides, nucleosides and bases by high pressure liquid chromatography. In: *Biological/Biomedical Applications of Liquid Chromatography* (ed. G. L. Hawk), pp. 295–331. Marcell Dekker Inc., New York & Basel

Carr, C. D. (1974) Use of a variable wavelength detector in high performance liquid chromatography. *Anal. Chem.* **46**, 743–746

Freed, C. R., and Asmus, P. A. (1979) Brain tissue and plasma assay of L-DOPA and α-methyldopa metabolites by high performance liquid chromatography with electrochemical detection. *J. Neurochem.* **32**, 163–168

Knox, J. H. (1976) Interaction of radial and axial dispersion in liquid chromatography in relation to the infinite diameter effect. *J. Chromatogr.* **122**, 129–145

Riggin, R. M., and Howard, C. C. (1979) Determination of benzidine, dichlorobenzidine, and diphenylhydrazine in aqueous media by high performance liquid chromatography. *Anal. Chem.* **51**, 210–214

Perrett, D. (1976) Simplified low pressure high resolution nucleotide analyses. *J. Chromatogr.* **124**, 187–196

Pippenger, C. E. (1979) High pressure liquid chromatography therapeutic drug monitoring: an overview. In: *Biological/Biomedical Applications of Liquid Chromatography* (ed. G. L. Hawk), pp. 495–506. Marcell Dekker Inc., New York & Basel

Savage, M. (1979) Mobile phase selectivity in the separation of PTH amino acids. *Altex Chromatogram* **2**, 1–2

Sivorinovsky, G. (1980) HPLC vs EMIT in therapeutic drug monitoring with emphasis on theophylline analysis. *Altex Chromatogram* **3**, 1–3

Warner, I. M., Callis, J. B., Davidson, E., Gouterman, M. P., and Christian, G. D. (1975) A two-dimensional rapid scanning fluorimeter. *Anal. Lett.* **8**, 665–669

Selected References for Further Reading

Deyl, Z., Macek, K., and Janak, J. (1975) *Liquid Column Chromatography; a Survey of Modern Techniques and Applications*. Elsevier, Amsterdam

Guiochon, G. (1980) Optimization in liquid chromatography. In: *High-performance Liquid Chromatography* (ed. C. Horvath), Vol. 2, pp. 1–56. Academic Press, New York

Hawk, G. L. (1979) (ed.) *Biological/Biomedical Applications of Liquid Chromatography*. Marcell Dekker Inc., New York & Basel

Johnson, E. L., and Stevenson, R. (1978) *Basic Liquid Chromatography*. Varian, California

Pryde, A., and Gilbert, M. T. (1979) *Applications of High-performance Liquid Chromatography*. Chapman & Hall, London

Yau, W. W., Kirkland, J. J., and Bly, D. D. (1979) *Modern Size-exclusion Liquid Chromatography*. John Wiley & Sons, New York

Biochemical Research Techniques
Edited by J. M. Wrigglesworth

5
Electron Microscopy

JOHN M. WRIGGLESWORTH

Department of Biochemistry, Chelsea College, University of London, London SW3 6LX, UK

5.1 Introduction

For best use of any microscope (light or electron) it is necessary to have a clear idea beforehand of what information is wanted from the specimen. Different types of microscopes have different advantages for any particular problem and it is important to choose the correct instrument to fit the information required.

It is important to remember that greater magnification and resolution do not necessarily lead to an increase in useful information. For example it would take a very patient investigator to use an electron microscope to study the branching of a nerve fibre inside a muscle. The amount of preparation and scanning of the specimen would be tedious in the extreme. A light microscope or even a magnifying glass could be used more quickly and provide a great deal more useful information. However, for detailed structural information at the sub-cellular level, the electron microscope is indispensable. Light diffraction limits the optical microscope in its ability to show details in objects smaller than one micron in size ($1\ \mu m = 10^{-6}\ m = 10^4$ Å).

Electron microscopy, or electron probe analysis as perhaps it should be called, can be used in a wide variety of ways for specimen analysis. The bombardment of a specimen by an electron beam provides much more information than a simple visualization of structure (Fig. 1). An electron beam hitting a sample can induce secondary electrons to be emitted from its surface as well as X-rays, Auger electrons and backscattered electrons. These can be analysed to give information on composition and topography. Absorbed electrons can be directly monitored by sensitive current measurement to provide similar information. Transmitted electrons can be visualized on a fluorescent screen to provide a 'conventional' picture but can also be analysed for

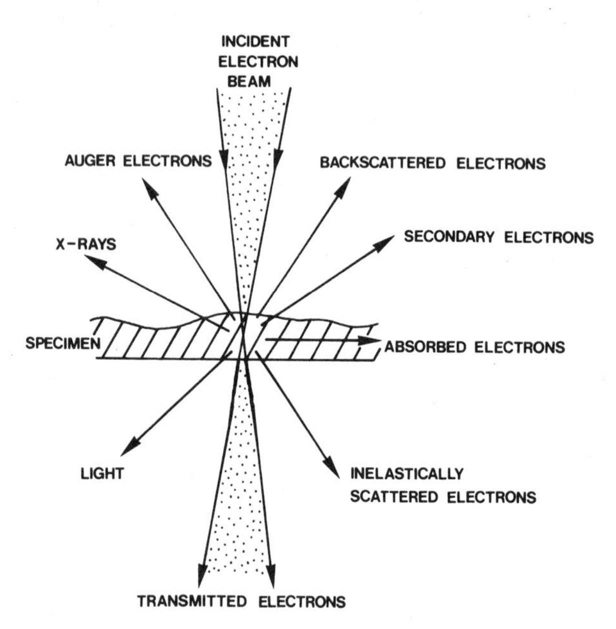

Fig. 1 Information possibilities in the electron microscope arising from various interactions of the primary electron beam with the specimen. Suitable detectors will allow analysis of specimen structure, composition, morphology and topography

energy and intensity by scanning transmission electron microscopy (STEM). One aim of this chapter is to try to help the reader evaluate the different electron probe techniques available, in order to decide on the most suitable approach to any particular problem should electron microscopy become necessary.

Finally, it must also be remembered that quite often the amount of useful information from a specimen is limited not by the resolving power of the microscope or even by practical considerations of time and cost but by the way in which the specimen has to be prepared for examination. This is especially true in electron microscopy where many preparative techniques are used. These include thin sectioning, negative staining, freeze fracture and replication methods. Each method provides information about the specimen in a slightly different form and each has its own special advantages. Once again it is important to choose the method which best suits the problem at hand.

5.2 Electron Probe Analysis Methods

5.2.1 Transmitted Electrons

5.2.1.1 Image Forming Effects

Electrons can be removed from the primary beam by scattering in two ways. Those that pass close to a nucleus of a specimen atom will be deflected through large angles. Because of the great difference in mass between electrons and nuclei, these elastically scattered electrons will suffer only a small loss of energy, analogous to a small ballbearing hitting a very much more massive sphere. The ballbearing is scattered through a large angle without imparting much energy to the larger sphere. However, because of the relatively small dimensions of atomic nuclei compared to interatomic distances, most incident electrons will be scattered by interaction with the electron cloud around the nucleus of the specimen atoms. These inelastically scattered electrons will have shared their energy with the specimen electrons and will suffer changes in both direction and energy. The probability of scattering for both elastic and non-elastic scattering will depend on the composition and thickness of the specimen, and on the energy of the primary electrons. An objective aperture can be used to remove a proportion of the scattered electrons and it is the loss of these electrons from the primary beam that eventually allows for the formation of an image in the transmission electron microscope. Making the aperture smaller will exclude more of the scattered electrons and contrast will be enhanced. Increasing the energy of the primary electrons by using higher accelerating voltages will reduce specimen interaction and, in this case, the final image contrast will be reduced.

The general principles of the transmission electron microscope are very similar to those of the optical microscope. The differences between the two instruments mainly result from the more complicated methods that have to be used to produce and focus a beam of electrons compared to the methods used for light rays. But once this can be done, 'electron optics' can be seen to be subject to the same rules and principles as light optics. In a simple optical microscope light rays from a lamp are condensed and focused on to the specimen by the condenser lens. The specimen is located just outside the focal point of the objective lens so that a real, magnified image is produced. This image then becomes the object for a second lens, the eyepiece, which turns it into a magnified, virtual image which is looked at by the eye, the final real image being formed on the retina. In an electron microscope (Fig. 2), beams of high velocity electrons are used instead of light rays. The source of electrons is a filament, usually made of tungsten, which is heated by a small current to incandescence. The filament (cathode) is kept at a high negative voltage with respect to a metal plate (anode) placed close by. The emitted electrons are attracted to the more positive anode and concentrated into a beam by having to pass through a small hole in a metal shield kept at the same negative potential as the filament. When it reaches the anode, part of the beam also passes through another small hole and continues down into the microscope. To prevent any collision of electrons in the beam with air molecules, the whole microscope, including the filament assembly, must be kept evacuated which normally means using pressures of less than about 10^{-6} torr (1 torr $\equiv$ 1 mm mercury $\equiv 1.32 \times 10^{-3}$ atmosphere). The beam of electrons is then condensed and focused on to the specimen by condenser lenses. Magnetic fields have to be used to do this since any lens made of solid material would just scatter the electrons. A magnetic field affects a moving electron by exerting a force on it at right angles to its direction of motion and at right angles to the direction of the magnetic field. This is the same principle as in the electric motor where a current carrying coil is induced to rotate in a magnetic field. To produce magnetic fields that will bend and focus a beam of electrons in a microscope, a very carefully designed magnetic lens has to be used. The focal length of such a lens can of course be altered at will by altering the current producing the magnetic field. The electron beam passing through the specimen is focused by an objective lens to form a magnified image which is then magnified further by other lenses. Instead of producing a virtual image which can only be viewed through an eyepiece, a projector lens forms a real image and focuses it on to a fluorescent screen. This is done because electrons cannot be seen directly by the eye and the image has to be projected on to a material which will fluoresce and give out a visible image. A photographic plate with an emulsion sensitive to electrons can easily be substituted for the fluorescent screen when a permanent record is wanted.

From a comparison of light optics with electron optics, we should expect that electron lenses would show aberration defects similar to glass lenses. This is

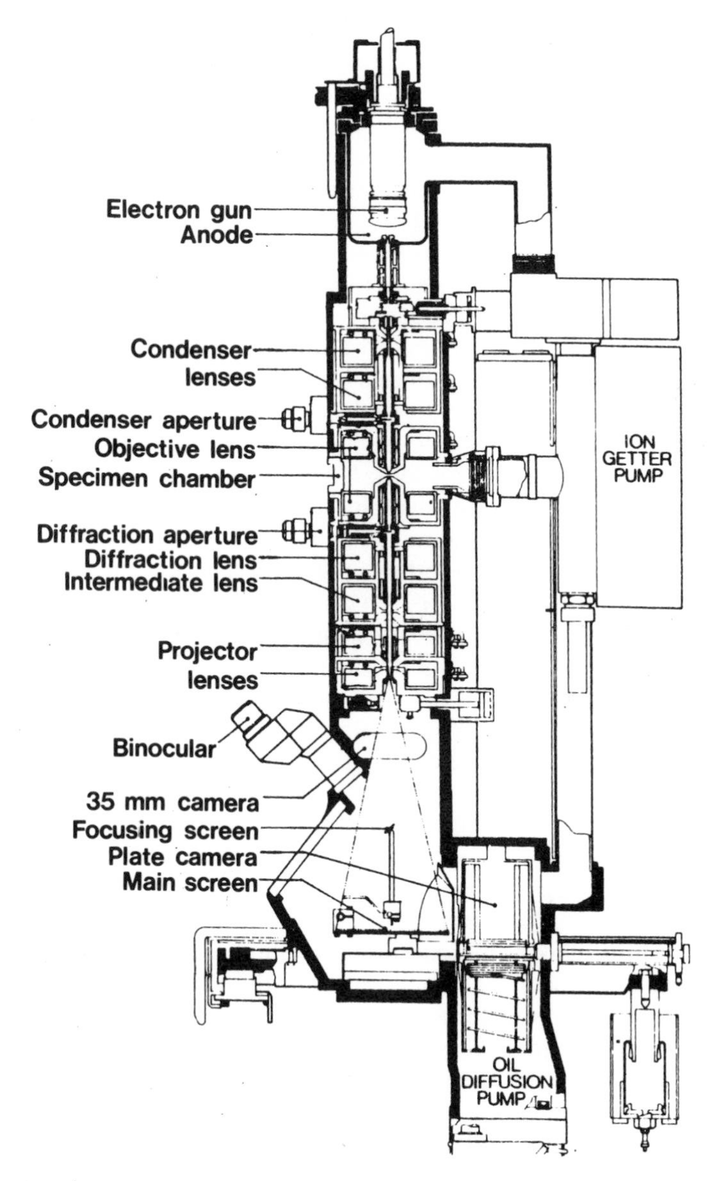

Fig. 2 The transmission electron microscope. Magnetic lenses are used to obtain an image of the specimen on a fluorescent screen. Magnification is geometric and subject to lens aberrations and imperfections. A high vacuum is necessary to reduce electron scattering in the column. (Reproduced by permission of N. V. Philips' Gloeilampenfabrieken)

found to be so. A condition is found in electron optics analogous to chromatic aberration in glass lenses where the focal length of the lens depends on the wavelength of the light. Not all the electrons in the beam move at exactly the same velocity since slight variations in the high voltage supply to the filament contribute to a slight spread in the accelerating force on the electron. Random collisions of electrons with any remaining air molecules in the 'vacuum' also cause a spread in velocities. These slight differences in velocity between electrons mean that the induced force on each electron in the magnetic lens is not the same. The point of focus of the beam will therefore be slightly blurred, the faster moving electrons coming to a focus slightly in front of the slower ones. A very stable high voltage supply is crucial in the electron microscope to reduce this 'chromatic' aberration to a minimum. Spherical aberration also occurs in electron lenses. This means that electrons far from the axis have a different focal point to those close to the axis. In light microscopy it is possible to reduce spherical aberration to negligible proportions by careful choice of radii for the two surfaces of the lens. Unfortunately this is not yet possible in magnetic lenses, and instead small apertures are used to limit the beam to around the central axis. These cannot be made too small in diameter without losing detail in the image so a certain amount of spherical aberration is always present in the electron microscope. In addition, problems of keeping magnetic lens currents constant and of controlling vibrations and stray magnetic and electric fields, all combine to limit the amount of detail seen in a specimen.

The contrast of fine details in a specimen can be enhanced by techniques of transmission electron microscopy analogous to those used in light optics. For example, high quality dark field images can be produced by insertion of a ring shaped aperture in the condenser lens which will only allow a hollow cone of electrons to fall on the specimen. Transmitted electrons that do not interact with the specimen can then be removed by a matching objective aperture. The formation of the image will then only be dependent on the scattered electrons. A bright image will appear against a dark background and the contrast will be very high. This technique is especially useful for the study of fine detail in small particles such as viruses, macromolecules and cell components.

A completely different approach can be used to produce a final image from the transmitted electrons. This is by the use of scanning transmission electron microscopy (STEM). In the STEM (Fig. 3), a beam of electrons is focused down through a condenser system to a small spot which is then scanned across the specimen in a Raster fashion, in a similar manner to the scan of an electron beam across a cathode ray tube. The transmitted electrons are captured by detectors beneath the specimen and the number of electrons (current) is amplified. The signal is connected to a cathode ray display tube which scans in synchrony with the beam scanning the specimen and modulates the brightness of the spot on the display tube. An image of the specimen can then be formed. The resolution of fine detail depends on the spot size of the scanning beam

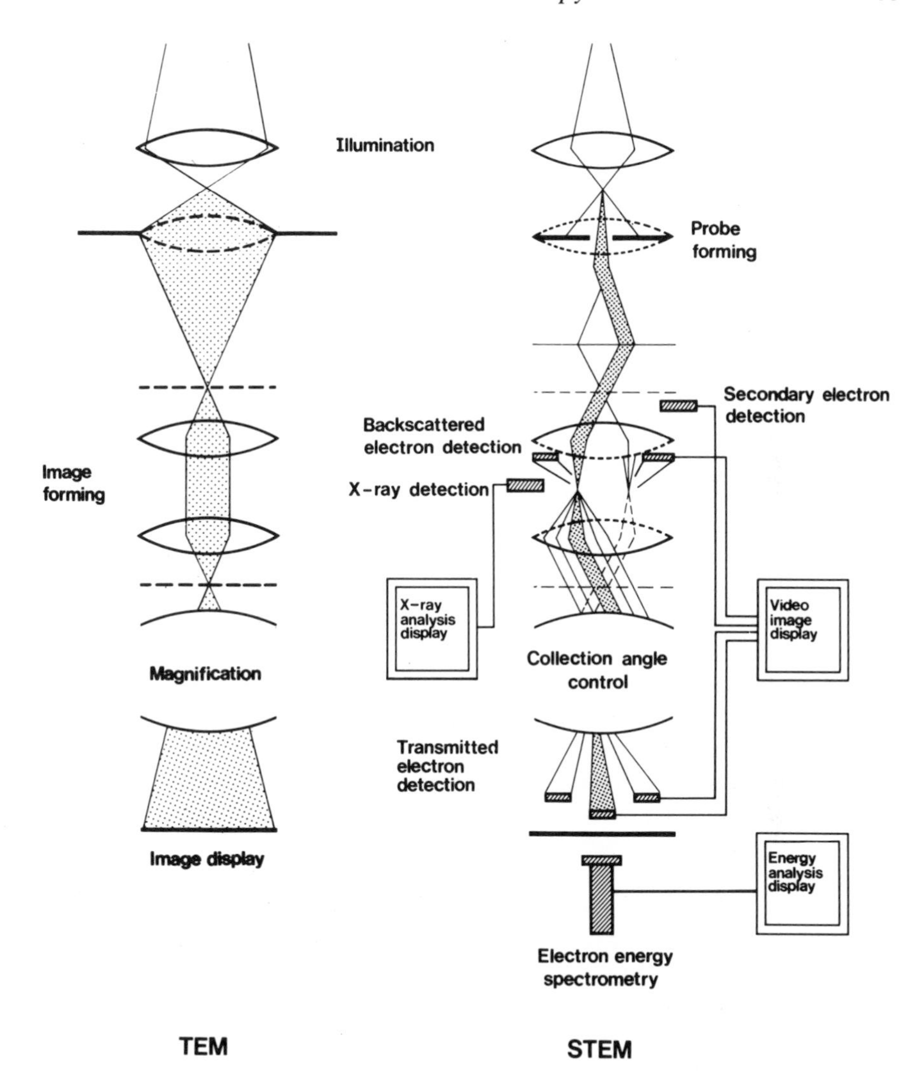

Fig. 3 Comparison of lens systems for the transmission (TEM) and the transmission scanning electron microscope (STEM). The magnetic lenses and the electron beam geometry are represented in the manner of a conventional optical ray diagram for ease of interpretation. The transmission scanning electron microscope differs from the straightforward geometric image-forming transmission microscope by producing a fine beam of electrons that can be scanned across the specimen. Transmitted electrons are captured and the signal (current of electrons) connected to a video image display system that scans in synchrony with the electron beam. Magnification is the ratio of the length of a single scan line on the specimen to the length of the imaging line on the video display, and is produced electronically without the use of imaging lenses. Various detectors can be positioned in the microscope to obtain information from the different interactions illustrated in Fig. 1. (Reproduced by permission of N. V. Philips' Gloeilampenfabrieken)

(which can be brought down to less than 1 nm diameter) and the magnification can be controlled by the relationship between the specimen area scanned and the size of the display tube.

The usefulness of the STEM lies in its versatility for analysis of the various interactions taking place between the specimen and the primary beam of electrons. As can be seen in Figure 3 various detectors can be fitted to analyse secondary electrons, X-rays, backscattered electrons and the energy distribution of the transmitted electrons.

5.2.1.2 Diffraction Analysis

When a specimen is made up of a regularly spaced array of structures, the primary electron beam containing electrons of the same energy (wavelength) will be scattered in a regular manner from each of the repeating structures. The inelastically scattered electrons (reflected rays) will interfere with each other and form a diffraction pattern which can be focused by a suitable lens system in a transmission electron microscope on to the final screen as a pattern of rings or spots. The spacings of the pattern will be related to the dimensions of the repeating structures in the specimen. Wide spacing in the pattern will arise from closely separated lattice structures.

Amorphous specimens with no regular repeating structures will not produce a diffraction pattern, and most biological specimens fall into this category. Those that do, such as a packed array of viruses or a crystallized preparation of enzyme, usually have large (>10 nm) repeat spacings and these used to be difficult to visualize in most microscopes. However, recent developments in lens design have made high dispersion electron diffraction readily available on many modern microscopes and the application of electron diffraction techniques to the life sciences should increase.

A diffraction technique that has proved very useful in the life sciences is optical diffraction analysis of the final photographed image of the specimen. This can reveal symmetry relationships which are not immediately obvious on study of the micrograph (Agar and Keown, 1978). A laser is used to provide a source of coherent light which is focused on to selected areas of the electron micrograph negative. The optical transform of the micrograph is then imaged by a series of glass lenses to produce the diffraction pattern. An especially powerful technique is to place a cardboard mask over the diffraction pattern with holes in it to let the diffraction spots pass through but which will cut out any other random rays of light. The diffraction rays can then be reconstituted to an image of the original micrograph which has enhanced contrast of the repeating structures and in which any 'noise' in the original micrograph is reduced (Fig. 4). Reconstitution can also reduce the effects of astigmatism and defocus in micrographs.

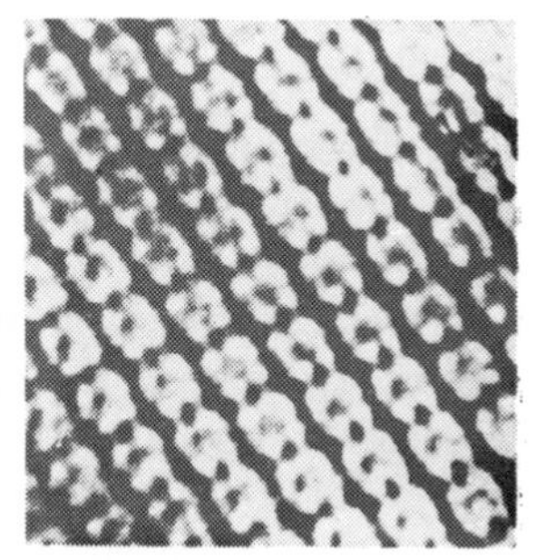

Fig. 4 Optical transform of a micrograph of phosphorylase b molecules. Left to right: a crystalline array of the enzyme molecules; the optical diffraction pattern from the micrograph shown left; the optically filtered image reconstructed from the main diffraction spots of the diffraction pattern (magnification 620,000 ×). The image is orientated with respect to its diffraction pattern shown in the middle picture. (Reproduced with permission from Eagles and Johnson (1972) *J. Mol. Biol. 64*, 693–695. Copyright: Academic Press Inc. (London) Ltd)

5.2.1.3 Energy Loss Analysis

An analysis of the energy of the scattered electrons passing through the specimen can provide information on the composition of the specimen. Inelastic scattering of the primary beam electrons will lead to energy loss by excitation of valence electrons and also by complete ionization of the specimen atoms. It is this latter interaction which is characteristic of the atom being ionized and thus composition details can be obtained about the specimen. Since the cross-section for ionization increases as the atomic number falls, electron energy loss analysis is especially useful for detection of elements of low atomic number ($Z < 12$) where X-ray microanalysis is limited (Isaacson and Johnson, 1975).

A STEM is normally used for energy loss electron spectrometry (Fig. 3). The inelastic electrons are allowed to pass through the column into an energy spectrometer. This is simply a sector-shaped magnetic field which spreads the narrow beam of electrons into a spectrum of velocities according to their energies. When this energy analysis is used in conjunction with the signal collected from elastic scattered electrons (backscattered electron detector in Fig. 3), an image with 'atomic number contrast' can be formed. This is because elastic scattering is approximately proportional (depending on the thickness of the sample) to an element's atomic number, whilst inelastic scattering is approximately proportional to the cube root of the atomic number. The signals from the two detectors can therefore be subtracted from one another electronically to produce a final image on the display tube which has enhanced contrast for the higher atomic number elements.

5.2.2 Backscattered and Secondary Electrons

The detection of backscattered and secondary electrons, and the formation of an image from the signals, is one of the most important electron probe methods for the investigation of surface topography. Secondary electrons are orbital electrons ejected from the specimen by interaction with the primary electron beam. They are of low energy when compared with the backscattered primary electrons and have low penetration in solid specimens. Those that escape mainly arise from interactions in the surface layer to a depth of around 10 nm for organic material and rather less for specimens coated with a heavy metal. The formation of an image is achieved by scanning electron microscopy.

Fig. 5 illustrates the basic construction of a conventional scanning electron microscope (SEM). This instrument enables a fine beam of primary electrons to be scanned across the specimen in the same way as for scanning transmission microscopy. The secondary electrons are attracted to a solid scintillator by a positively charged grid. The electrons hitting the scintillator induce flashes of

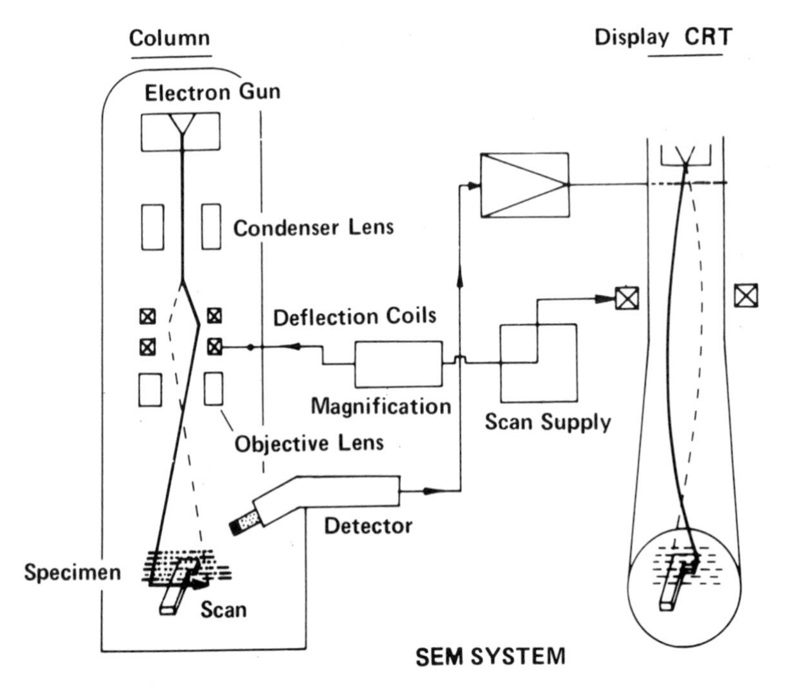

Fig. 5 Basic construction of a scanning electron microscope. Secondary electrons from the surface of the sample are collected to provide an electrical signal which is used to modulate the brightness of the spot on a video display. The cathode ray tube beam is scanned in synchrony with the electron beam moving across the specimen and a surface image can be constructed. Reproduced by permission of Perkin-Elmer Ltd

light which can be amplified in a photomultiplier, and the final electrical signal is then used to modulate the brightness of the display spot on a cathode ray tube. The cathode ray tube beam is scanned in synchrony with the electron beam across the specimen and thus a surface image can be constructed.

Backscattered electrons can be collected by the same detector but because of their higher energies they have approximately straight trajectories to the detector, and signal yield is best improved using a separate detector with a large solid angle (Reimer, 1978).

With non-conducting specimens, secondary electrons unable to penetrate to the surface and escape are absorbed in the specimen and can build up local charge areas. The strong electrostatic fields caused by charging the specimen can distort the primary electron beam and hence the final image. To minimize this effect the sample is quite often coated with a fine deposit of a conducting metal such as gold. For biological specimens, there is also the possibility of making the sample conducting by chemical fixation techniques (Kelley *et al.*, 1973).

Solid inorganic materials can be examined with minimum specimen preparation but most biological samples usually have to be dehydrated. Artefacts of drying can be minimized by critical point drying but this still requires substitution of the water by intermediate fluids, and fixation may be necessary. Some biological specimens with thick surface coats, such as various insects or plants, may be examined directly in the hydrated state, as can frozen hydrated material if a suitable cold specimen stage is present.

The resolution limit of a SEM depends directly on the size of the scanning spot of the electron beam, and since a high primary electron intensity is required to produce sufficient secondary electron emission, the spot size is usually an order of magnitude greater than that for the STEM. Nevertheless, resolution of detail down to less than 10 nm can be achieved with field emission sources in the electron gun. The greatest advantage of scanning over transmission microscopy is the huge increase in surface information obtained. One only has to look at the surface detail in specimens such as the cultured cells shown in Chapter 7 (p. 212) to appreciate the usefulness of SEM in the biological sciences.

5.2.3 X-Ray, Auger Electron and Cathodoluminescence Analysis

When the orbital electron of a specimen atom is excited and ejected by a primary beam electron, the resulting orbital vacancy is filled by other specimen electrons falling from higher energy orbitals. The energy released in this process can be emitted as electromagnetic radiation (X-rays) or can be used to excite and eject another high orbital electron (Auger electron). In some materials, transfer of electrons from conduction bands to the secondary vacancies in outer orbital bands may also occur. The energy released in this case will be in the form of light (cathodoluminescence) or may even be converted to vibrational energy (heat) in the sample. These processes are illustrated in Fig. 6. With appropriate detector systems, each can be utilized to provide useful information about the specimen.

The energy spectrum of X-rays from a sample irradiated with high energy electrons comprises a continuous, non-specific background ('Bremsstrahlungen') in which can be seen specific lines from the elements present. These specific lines can be detected by dispersing the radiation into a wavelength spectrum using a crystal as a diffraction grating. Specific emissions are focused on to a detector, a different crystal usually being used for each element measured. This is the 'wavelength dispersive' method and has the advantage of being capable of separating narrow lines at low intensities. The alternative, 'energy dispersive' method, has the advantage of analysing all the X-ray spectrum at one time with a semiconductor detector whose response is proportional to the energies of the incident X-rays. The non-specific background is also collected, reducing the signal-to-noise level compared to the wavelength dispersive spectrometer, but sensitivity is high since the detector can be positioned close to the specimen to increase the solid angle for X-ray collection (Reed, 1975; Statham, 1981).

X-Ray microanalysis can be operated in both fixed or scan modes. In the fixed mode, the primary beam is focused on to the particular area of the specimen of interest and an elemental analysis generated. Alternatively the beam can be scanned across the specimen and a distribution map of the elements produced. The method is now finding considerable application to the biological

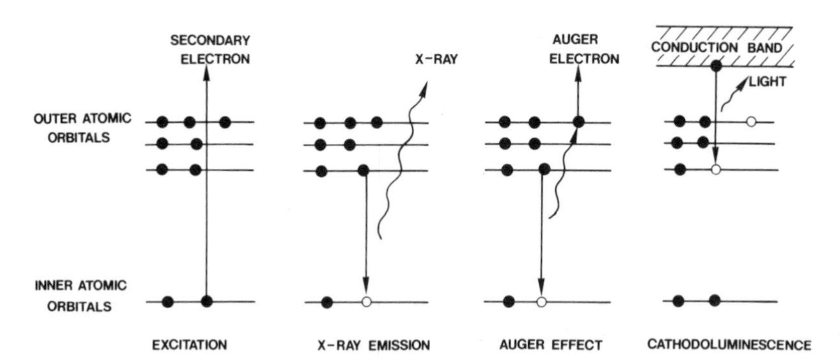

Fig. 6 Production of secondary electrons, X-rays, Auger electrons and cathodoluminescence by electron irradiation. Secondary electrons are excited and ejected from the inner orbitals of specimen atoms by interaction with the primary electron beam. The energy released by electrons falling from higher energy orbitals into the resulting orbital vacancies can be emitted as X-rays or can be used to excite and eject another higher orbital electron (Auger effect). Cathodoluminescence results from the energy release by transfer of electrons from conduction bands in the specimen to vacancies in the outer orbitals

sciences (Hall, 1979), especially when frozen-hydrated sections are used (Gupta and Hall, 1981).

X-Ray yield drops strongly for lower atomic number elements and below 10 the yield is usually beyond detection. However, the Auger effect rises for low atomic number elements and below atomic number 10 emission can be observed for all the elements (above helium) with almost the same probability. The short mean free path of Auger electrons in the sample means that only the surface layers can be analysed and quantitation can be difficult. In addition, the Auger electrons, with specific energies, have to be distinguished against a continuous background of backscattered and secondary electrons. Nevertheless useful information on elemental depth concentration profiles can be obtained (Holm and Reinfandt, 1977; Janssen and Venables, 1979).

Cathodoluminescence occurs as the primary electron beam causes the emission of light in the UV, visible and IR regions. Not all materials exhibit cathodoluminescence and the photon yield from organic substances is usually very much less than that for inorganic phosphors. Efficient photodetectors are therefore important. Biological samples have to be fixed and dehydrated before examination to prevent beam damage but frozen hydrated specimens can be used. As well as autoluminescence, many organic samples can be treated with specific fluorochrome dyes such as quinacrine dehydrochloride and acridine orange (Barnett *et al.*, 1975). An application that will no doubt increase in use is to couple the fluorescein label to specific antibodies which can then be located by SEM using a cathodoluminescence detector. For example, this approach has been used by Soni *et al.* (1975) to locate anti-IgG-induced cap areas on mouse B lymphocytes.

Finally, in some cases the number of electrons absorbed by the specimen can be measured directly by connection to a suitable current amplifier. The resulting signal can be used to modulate the brightness of the signal in the SEM. In conducting and semiconducting materials this can provide useful additional information on the 'mass thickness' of the specimen. Unfortunately biological applications are limited by the low conductance of most organic materials.

A comparison of various analytical methods of electron beam microscopy is given in Table 1.

5.3 Specimen Preparation Methods

As mentioned in the Section 5.1, specimen preparation is of crucial importance in electron microscopy. The right choice of preparation has to be matched not only to the method of analysis but also to the information wanted. Quite often there is no agreement between the appearance of the same specimen prepared for microscopy by different methods. The relationship between what is seen and what may be the biologically active structure in the cell is a matter of judgment. For example, disrupted mitochondria examined by negative stain-

Table 1 Analytical methods of electron probe microscopy

Analysis	EM	Information	Resolution	Reference
Energy loss	TEM/STEM	Mass thickness	10 nm	Isaacson and Johnson, 1975
Diffraction	TEM	Repeat spacings	<0.2 nm (~1000 nm area)	Schwartz and Cohen, 1977
X-ray	SEM/TEM/STEM	Elemental analysis ($Z > 10$)	10–100 nm	Hall, 1979
Auger effect	SEM	Surface elemental analysis ($Z > 2$)	100 nm	Janssen and Venables, 1979
Cathodoluminescence	SEM	Molecular composition and structure	500 nm	Herbst and Hoder, 1978
Direct current	SEM	Mass thickness (conducting materials)	10–100 nm	Newbury, 1976

ing clearly show 'knobs and stalks' protruding from the inner membrane, known to be associated with H^+-ATPase activity. These surface projections are not seen in etched frozen preparations (Wrigglesworth *et al.*, 1970) and cannot be resolved to any satisfaction in thin sections. Nevertheless, each of these preparation methods has contributed in its own way to a knowledge of mitochondrial structure. It is therefore important to appreciate the advantages and limitations of various preparation techniques before choosing the best electron microscopy approach to the problem at hand.

5.3.1 Thin Sections

The formation of the final image in the transmission electron microscope depends on various areas of specimen removing or scattering more electrons from the incident beam than others. This results in light and dark areas in the image that hopefully correspond to structures in the specimen. Absorbance and scattering depend on atomic composition and thickness. A thick, high-mass specimen will be strongly absorbing and no detail will be seen. As most biological materials are made up of the lighter elements (carbon, hydrogen, oxygen and nitrogen), it can be calculated that a thickness greater than 0.5 μm will give excess scattering under normal high voltages and no transmission detail will be seen. Usually biological section thickness is kept below 0.1 μm. This compares with specimens for light microscopy where 5–10 μm thick sections are used.

5.3.1.1 Histological Ultrathin Sections

Most cells fall in the size range 10–100 μm and many subcellular organelles are of the order of 1 μm. For ultrathin sections therefore these specimens have to be embedded in a polymerized resin, made from carbon and hydrogen to keep electron scattering to a minimum, and thin sections cut from the hardened block of material. The specimens have to be chemically treated to preserve molecular structure prior to embedding and sectioning, and enhancement of contrast by specific heavy metal staining may be necessary.

As can be imagined, many possible artefacts and distortions can arise during these procedures. The whole method is time consuming and requires great skill, but the results of the thin-sectioning technique have proved so successful that electron micrographs of material prepared in this way have given us the 'standard' view of biological structure.

Fig. 7 outlines the basic stages of thin section preparation. A small amount of sample is first placed in a chemical fixative to stabilize the structure. The size of the sample must be sufficiently small to allow free and quick internal diffusion of the fixative. A sample from tissue should not be more than a few mm thick, and in cutting this small amount from the tissue care must be taken to avoid

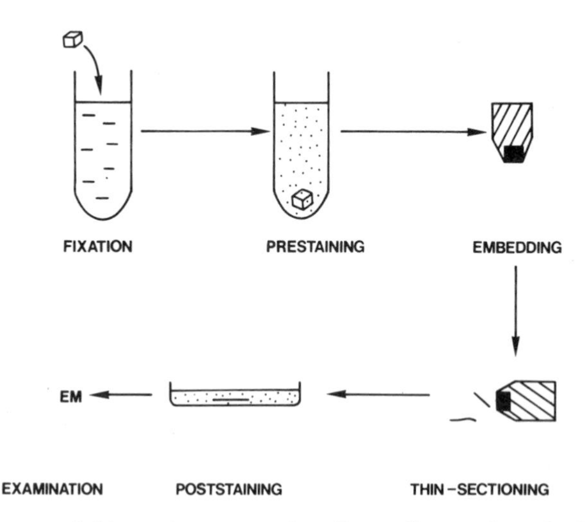

Fig. 7 Basic stages of thin section preparation. Depending on the embedding medium, dehydration of the fixed specimen may or may not be necessary before embedding and thin sectioning

damage. Particulate material can be centrifuged to a small pellet. Several chemical fixatives have been used in electron microscopy to preserve structure but nowadays the two most common are glutaraldehyde and osmium tetroxide. They are often used in conjunction as both have different modes of action which complement one another. Glutaraldehyde treatment is usually the first stage. The sample is placed in 1–2% fresh glutaraldehyde solution, buffered to the required pH (usually pH 7), together with some form of osmotic support (inpenetrant salt or sucrose) to preserve tissue ultrastructure. Glutaraldehyde reacts with free amino groups, mainly on proteins, cross-linking the proteins into a stable network (Molin *et al.*, 1978). Most of the fixation reactions with glutaraldehyde take place within a few seconds if diffusion through the sample is not rate limiting, which makes it a useful fixative for samples undergoing time-dependent structural changes such as occur during muscle contraction, mitochondrial swelling or phagocytosis. Particulate samples can be fixed in suspension before pelleting. The sample is treated with glutaraldehyde for around an hour at 0–4 °C to ensure full reaction and then washed in buffer before the addition of freshly made osmium tetroxide solution (1–2%) at pH 7. Specimens can be stored for several hours, or overnight if required, before osmium addition. Osmium reacts with double bonds in both proteins and lipids, forming cross-links and stabilizing structure (Stockenius and Molin, 1965). As a primary fixative and electron dense stain it can cause damage to subcellular structures, but when used after glutaraldehyde it successfully fixes and stains various cell components.

An alternative to osmium is to use 1% buffered tannic acid followed, after washing, by 3% methylamine tungstate solution. This treatment produces even more intense staining than osmium (Skaer, 1981) but there are permeability problems with tannic acid for some cell types.

Water-soluble molecules that have not reacted with either of the fixatives are lost from the sample but most macromolecular structures are preserved. Many enzymes retain slight activity especially if the fixation time is short, formaldehyde (0.5–1%) is substituted for glutaraldehyde, and osmium is not used. If required this can be made use of in later staining procedures with histochemical stains linked to specific enzyme activities.

Several materials have been developed for embedding the fixed material into a hard enough resin for thin sections to be cut. They fall into three general groups: epoxy, polyester and acrylic. Possibly the epoxy resins have found the widest use. These include Araldite (diglycidyl ether of *bis* phenol), Epon (triglycidyl ether of glycerol) and Spurr (vinyl cyclohexane dioxide). These resins are hardened, using anhydrides accelerated by amines, usually at 60 °C overnight. Various additives can be used to modify block hardness and stability of the resin under the electron beam. The main problem with some epoxy resins is their high viscosity, making tissue penetration difficult, especially for plant material, and their incompatibility with water, making it necessary to dehydrate the specimen before embedding. The fixed sample is usually dehydrated by successive washes in aqueous alcohol solutions of decreasing water content. Any unfixed non-polar molecules are lost at this stage, for example free saturated fatty acids. Specimens that have to be stored for a few hours, or overnight, are best left in 70% ethanol in water at 4 °C. Embedding is usually carried out in two stages, the first using 50% resin diluted with a non-polar solvent such as propylene oxide to ensure good penetration and then finally using the viscous 100% resin. After block hardening, thin sections can be cut on a high precision microtome. An unfortunate property of epoxy resins is that they do not completely polymerize and, on hot days, section cutting can be difficult due to creep. However, the main advantage of epoxy resins is that they generally harden uniformly with little shrinkage (less than 5%) and are stable in the electron beam.

Polyesters used for microscopy (e.g. Vestopal) usually comprise a mixture of three components: an unsaturated low molecular weight polyester resin, a fluid monomer capable of radical polymerization and a heat activated catalyst. Radical production can be induced by heat, chemical activation or light. Oxygen inhibits the reactions and hence air should be excluded wherever possible. Hardness is controlled by the degree of cross-linking which in turn is controlled by the concentration of the cross-linking monomer. Again dehydration of the specimen is necessary before infiltration with the resin, but the sections are very stable under the electron beam.

All acrylic monomers have the acryl group in common and are polymerized by radical chain reaction. Once again air should be excluded because of the competition of oxygen for the radical on the growing polymer chain. A large choice of acrylics is now available including water-soluble mixtures (e.g. glycol and hydroxypropyl methacrylates) which enable embedding to occur without prior dehydration of the specimen. The main advantage of the acrylics is that they are often of low viscosity giving good infiltration.

The choice of embedding material for any particular specimen is best made in the light of previous experience and several reviews are available to give guidance (see for example Hayat, 1981).

Thin sections are cut from the hardened material on an ultramicrotome. Often thicker sections are first prepared for light microscopy examination to ensure that the required area of sample is being sectioned. Finding small samples such as individual cells for sectioning is often a problem in electron microscopy and it may be necessary to embed the fixed material in paraffin or agar and examine the samples under phase contrast. The particular cells can then be selected for embedding and sectioning in the normal way.

Post-staining of the thin sections, once mounted on the specimen grid, is an important procedure to enhance contrast of particular structures. A wide variety of stains have been used (Table 2) but all depend on the presence of a heavy metal to provide 'electron contrast' in the microscope (Lewis and Knight, 1977). Osmium provides a certain amount of electron dense stain but further contrast and detail can be produced by treatment with a staining solution of lead, tungsten or uranium salts. These salts react with a number of cellular components including nucleic acids (Table 2). In general, however, the staining is not particularly specific and various attempts have been made to induce specificity in staining for electron microscopy. Several specific enzyme stains have now been developed (see the series of volumes by Hayat, 1973–1977). Most depend on heavy metal capture by the product of the enzyme reaction. The technique with phosphatases for example is to incubate lightly fixed thick sections in a medium which contains the specific enzyme substrate plus a soluble lead salt such as lead nitrate. Deposition of lead phosphate then occurs at sites of phosphatase activity. The thick sections are washed (avoiding phosphate buffers!), fixed and lightly stained with osmium prior to embedding and thin sectioning. The heavy metal capture technique has been extended to various soluble ions such as in the ultrastructural localization of potassium by tetraphenyl boron precipitation (Van Steveninck and Van Steveninck, 1981).

Autoradiography at the electron microscope level provides a most useful technique for localizing various cellular reactions. The tissue is specifically labelled with an alpha-, or low energy beta-emitter and ultrathin sections prepared in the normal way. Photographic emulsion is applied to the sections which are then stored (in lightproof conditions) for a period of time. The radiation from the localized isotope activates the silver halide crystals in the emulsion and these can be developed to form electron dense silver deposits on

Table 2 Some heavy metal stains commonly used for post-staining ultrathin sections

Metal	Stain	Specificity	Comments	Reference
Lead	Alkaline lead hydroxide	General	Contaminating lead carbonate precipitation in presence of CO_2	Karnovsky, 1961
	Lead–ammonium acetate	General	Less staining density than lead hydroxide	Björkman and Hellström, 1965
	Lead citrate	General	Simple and reproducible	Reynolds, 1963
Uranium	Uranyl acetate	Anionic groups e.g. DNA, phospholipids	Low solubility in water	Hayat, 1981
Tungsten	Phosphotungstic acid	Glycoproteins	Avoid osmium in fixation process if specificity required	Gordon and Bensch, 1968
Silver	Ammoniacal silver	Nucleoproteins	Oxidizing stain, will react with aldehydes	Black and Ansley, 1966
	Silver nitrate	Lipofuscin pigment		Hendy, 1971
Iron	Colloidal iron	Membrane sialic acids	Poor penetration into sections	Mowry, 1958
Chromium	Gallocyanin–chrome alum	Nucleic acids	Staining can be quantitative	Sandritter *et al.*, 1981

the section. The application of this technique to enzyme location is described in detail by Jacob and Budd (1975).

Perhaps the most interesting developments in stain specificity lie in immunochemical methods based on specific antibody binding (Sternberger, 1979). The basic technique is to raise and purify monospecific antibodies to the protein under study. These can be prepared by injection of the purified protein into rabbits. The lightly fixed samples are incubated with these antibodies followed by incubation with anti-rabbit antibodies prepared from another species, for example goat. The anti-rabbit antibodies have been conjugated either with a heavy metal derivative, such as colloidal gold (Gueze *et al.*, 1981) or with an enzyme which can be stained specifically as mentioned above. The classical enzyme conjugate is horseradish peroxidase (Sternberger *et al.*, 1970). Further fixation followed by thin section preparation and staining then follows.

An annoying effect with many heavy metal stains, lead in particular, is that insoluble precipitates tend to form on the sections on exposure to air whilst drying. These can be avoided if exposure to air or carbon dioxide is prevented, but there are some useful tricks to help remove stain precipitates. Oxalic acid solutions can be used to remove precipitates of uranyl acetate (Avery and Ellis, 1978) and 10% glacial acetic acid can be used to remove stain precipitates resulting from either lead salt treatment or double staining with uranyl acetate and lead citrate (Kuo, 1980).

Finally it is important to remember that many of the materials routinely used in the preparation of ultrathin sections are highly toxic. Osmium in particular should be handled with great care. Small ampoules should be used, wherever possible, in a high containment fume cupboard. Any osmium spills can be immediately inactivated by a reducing agent such as powdered sodium ascorbate. The black product can then be easily located and removed. Many of the ingredients of resin mixtures are hazardous (Causton, 1981) and users should carefully read any instructions accompanying the materials.

5.3.1.2 Frozen Hydrated Sections

A wide variety of equipment has been used to prepare and examine frozen hydrated specimens (see Robards and Crosby, 1979, for review). By this method, the chemical treatments of the previously described sectioning methods can be avoided (although most specimens are lightly fixed prior to freezing). With the increased use of X-ray microanalysis, cryosectioning is assuming increased importance.

The major problems lie in ensuring that the specimen is never exposed to conditions that may cause it to melt or sublime. Rapid freezing of the tissue by immersion in a liquid nitrogen slurry will retain the position of diffusible ions so that electrolyte concentrations can be measured subsequently. If required, the tissue can be pretreated with a cryoprotective agent such as dimethylsulphoxide.

Contamination of the section by ice crystals has to be avoided by vacuum or moisture-free gas systems. It has been argued that the process of sectioning frozen material causes local melting at the cutting site, the mechanical energy of the knife edge being converted to heat which warms the cut section. However, the results of cryosectioning when compared to freeze fracture of the same replicas do not indicate that appreciable melting changes take place (Frederik and Busing, 1980) and displacement of diffusable ions and other soluble compounds is likely to be negligible compared to the current resolving power of X-ray microanalysis.

5.3.2 Negative Staining

Negative staining is a far simpler technique than thin sectioning. It does, however, have its limitations. The basis of the technique is to examine the specimen in the presence of a non-interacting electron dense stain against which the sample stands out in outline (Fig. 8). A 'negative' image (light against a dark background) is seen. The technique has little application to intact tissue but is ideal for studying the outline structure of small pure materials such as viruses or large proteins.

The stain is usually a water-soluble heavy metal salt such as phosphotungstic acid, made up in a 1–2% solution at neutral pH. A drop of this staining solution is mixed with a drop of the sample and the mixture spread on the support film of an electron microscope grid. Thin support films are commonly prepared from nitrocellulose (collodion), polyvinyl formal (Formvar) or carbon. These have low mass thickness and good support strength. Sample spreading can be achieved by placing a drop of the stain and sample mixture on the grid and gently drawing away excess liquid with filter paper. Alternatively the mixture can be sprayed directly on to the grid. This is often the method of choice when aggregation of the sample has to be avoided. Most support films, especially Formvar and carbon, are very hydrophobic and sample materials in aqueous

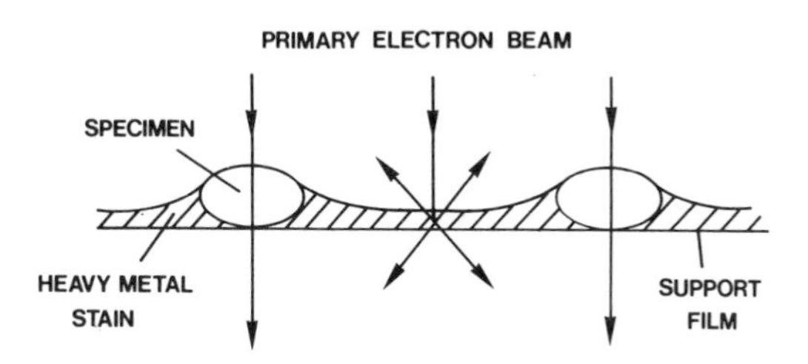

Fig. 8 Basic principle of negative staining. The specimen is spread on a coated grid in the presence of a non-penetrating heavy metal salt solution. After drying, the grid can be examined in the transmission microscope. Electrons hitting the areas of electron dense stain surrounding the sample are strongly scattered, providing a dark background for the lighter sample image

suspension may aggregate. If it is possible to add serum albumin (1%) to the suspension without affecting the sample, the spreading properties of the suspension on the support film can be enhanced. Once dry, the grid can be examined in the microscope. Electrons hitting the areas of electron dense stain surrounding the object are strongly scattered, providing a dark background for the lighter sample image (Fig. 9). Any penetration of stain into the specimen will outline internal structure. Sucrose should be avoided in sample suspensions since on drying it can interfere with the evenness of stain penetration. Some stains will interact with specific material in the sample and positive staining may occur, for example phosphotungstate staining of the glycoproteins on bacterial cell walls. Details of various stains and negative staining techniques can be found in Haschemeyer and Meyers (1972).

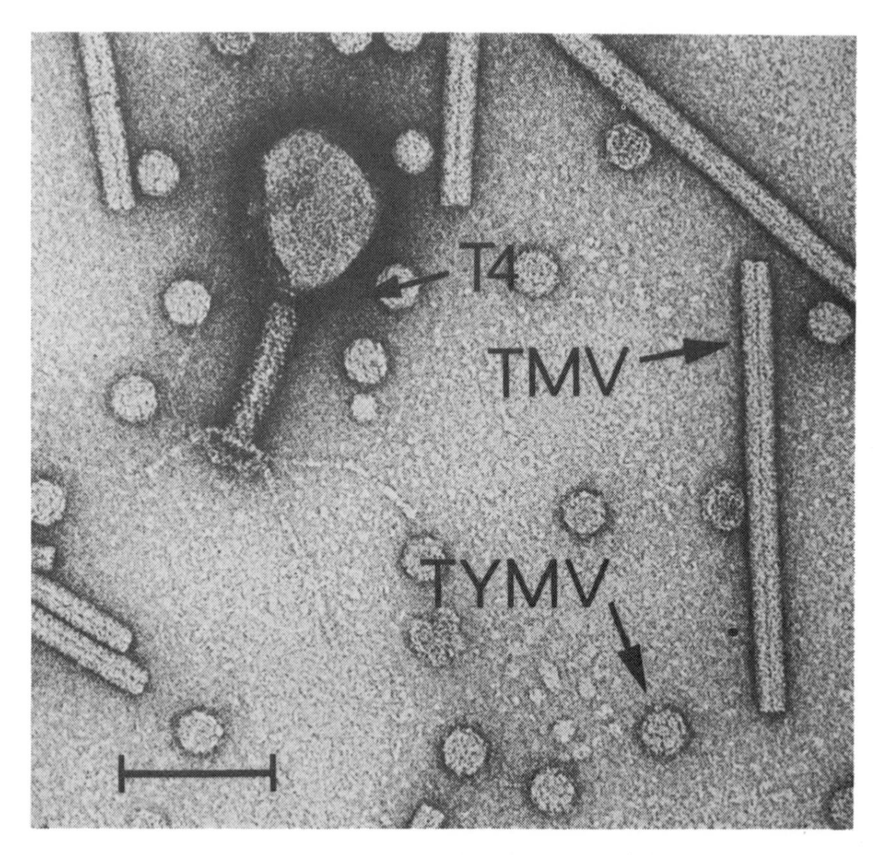

Fig. 9 Electron micrograph of negatively stained viruses. Bacteriophage (T4), turnip yellow mosaic virus (TYMV) and tobacco mosaic virus (TMV) are shown after negative staining with 1% uranyl acetate solution. Bar = 100 nm. (Micrograph courtesy of J. T. Finch)

5.3.3 Heavy Metal Shadowing

Another technique for outlining the structure of an object in the electron beam is to 'shadow' it with a heavy metal (Fig. 10). A thin wire of the metal (for example platinum or gold) is heated in vacuum at the tip of a small carbon arc until evaporation of the metal occurs. The evaporation chamber is placed at an angle to the specimen spread on a grid so that directional deposition occurs. The method is particularly suited to bacteria and viruses because of their size and stability under shadowing conditions. Resolution by shadowing is limited by the thickness of the layer of evaporated metal and is therefore usually less than with the negative staining technique. Fine detail can easily be lost as the layer of metal builds up. However, the technique has proved especially useful for nucleic acid examination as shown in Fig. 11.

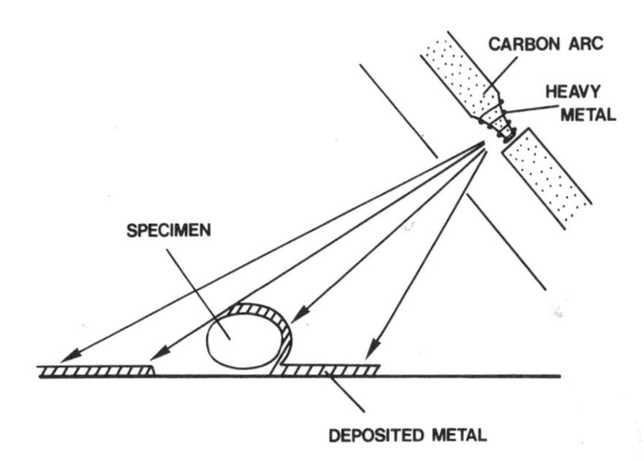

Fig. 10 Splutter shadowing of a specimen with a heavy metal. Evaporation of the metal occurs at the tip of a small carbon arc. Angled deposition in vacuum provides a replicated surface for examination by scanning or transmission microscopy

5.3.4 Freeze-fracture

The freeze-fracture technique can provide invaluable information on intracellular topology. Hydrophobic or apolar interactions, being strongly entropy dependent, weaken as temperature is lowered. This means that when a sample of frozen biological material is fractured by a sharp blow, quite often the fracture plane will follow the hydrophobic line along the centre of lipid membranes. The fracture face exposed can be replicated to provide topological information on membrane organization and will also reveal information on the molecular organization of the membranes themselves. The basic stages in the technique are shown in Fig. 12.

Artefacts caused by ice crystal growth during specimen freezing represent one of the main problems of the technique. Severe mechanical damage is

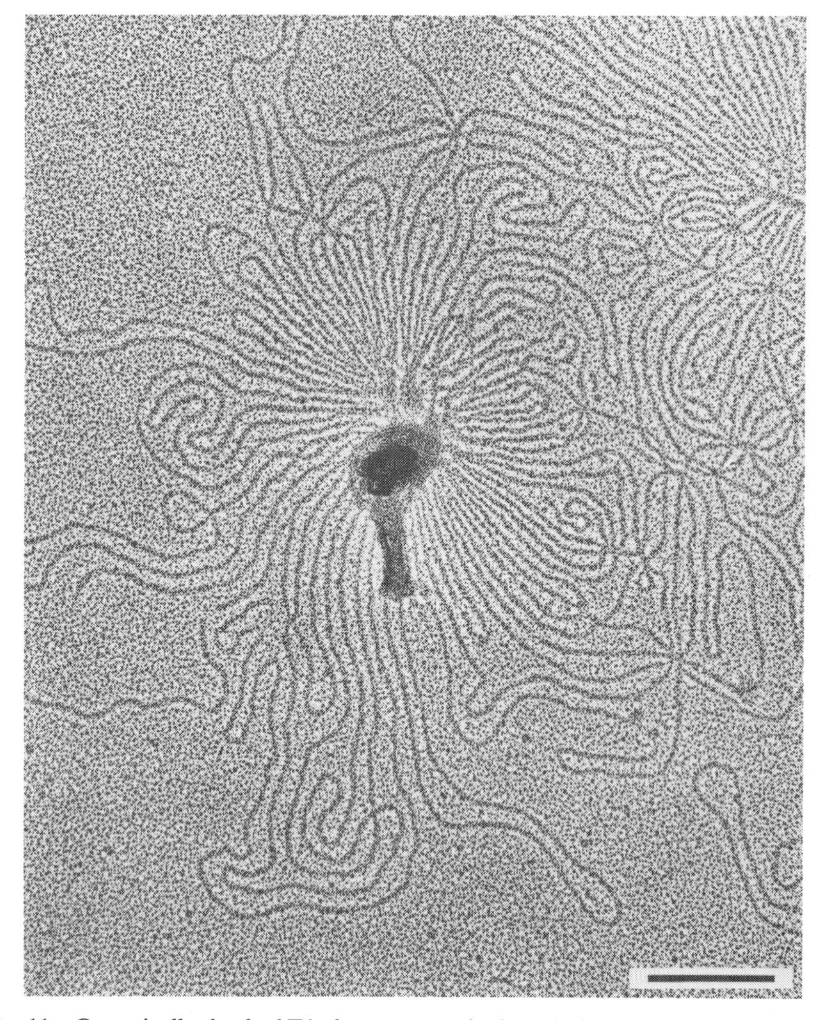

Fig. 11 Osmotically shocked T4-phage, rotary shadowed with platinum to show expelled nucleic acid. Bar = 200 nm. (Micrograph courtesy of Dr. W. Tichelaar, State University of Groningen, The Netherlands)

caused to intracellular structures by ice crystal formation and the concentration and distribution of water-soluble materials changes as the intracellular water freezes, excluding ions and other water-soluble molecules. An approach to minimize these problems is to infiltrate a cryoprotectant such as glycerol (20–50%) into the material before freezing. This is thought to reduce the amount of free water available for ice crystal formation. A eutectic mix forms on cooling which will solidify into a homogeneous background. Unfortunately

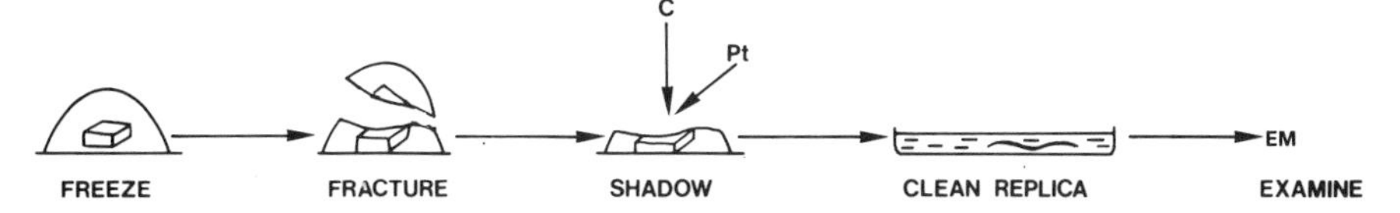

Fig. 12 Basic stages of freeze-fracture specimen preparation. Fracture, etching and replication are carried out under vacuum

artefacts of cryoprotection have long been recognized even when glutaraldehyde prefixation is used (for a review see Rash and Hudson, 1979). Perhaps the best approach is to attempt to reduce ice crystal size (to 10 nm or less) by fast freezing so that aldehyde prefixation and glycerol impregnation can be avoided.

Immersing samples into liquid nitrogen is not a particularly 'fast' freezing method. Under normal pressure liquid nitrogen exists between −210 °C to −196 °C (63–77 K), a very narrow range. When a sample is placed in the cold liquid, a surface layer of nitrogen gas soon forms around the material. Nitrogen gas has low thermal conductivity and cooling of the specimen is then impeded. An alternative is to immerse the sample in liquid Freon-22 or liquid propane, both cooled by liquid nitrogen, where this particular problem can be minimized. Small specimens can be placed into liquid nitrogen slurry (at −210 °C) formed by placing liquid nitrogen under vacuum for a few minutes, the evaporation causing cooling to the nitrogen freezing point. Very fast freezing rates can be achieved by liquid propane jet methods. The sample is sandwiched in a hat-shaped specimen support (usually made of thin copper) and a liquid propane jet is forced on to one or both sides. The fast cooling rates of sandwiched specimens (shown in Table 3) will be strongly affected by specimen mass. A value of around 2 *W*/m K for the thermal conductivity of ice (Ratcliffe, 1962) gives cooling rates of around 50 μm depth per msec. Sample sizes greater than a few mm radius should be avoided.

Table 3 Cooling rates of 50 μm thick sandwiched samples of cell culture medium frozen by different methods (data from Pscheid *et al.*, 1981)

Mode of freezing	Cooling rate ($K\ sec^{-1}$)
Dipping into N_2 'slush'	800
Dipping into Freon-22	1350
Dipping into liquid propane	2400
Propane jet	3100

Once frozen the sample is transferred to the fracture apparatus. Different types of equipment are available for fracture and replication, some requiring a large capital outlay, but essentially two methods of fracture are used. The first employs a knife which scrapes rather than cuts the sample. The second method is particularly suited to frozen suspensions. The sample mixture is frozen inside two small hollow copper holders placed one on top of the other. A sharp knock will separate the holders and expose the fracture face of the specimen. In both methods the facture process is done under high vacuum to avoid surface contamination and usually at temperatures lower than −100 °C to avoid ice sublimation. Once fractured, the specimen temperature can be raised to around −100 °C for ice-sublimation ('etching') to occur. This can reveal surface structure information in addition to fracture face detail. The fractured surface is then replicated usually by platinum, followed by carbon for replica stability, and removed from the fracture equipment for cleaning and examination. A series of sodium hypochlorite solutions (30–70%) will remove most biological material from replicas. Further cleaning in warm 50% saturated sodium hydroxide solution can be carried out if required.

Fig. 13 shows the appearance of a sample of erythrocyte membranes examined by the freeze-fracture technique. Cross-section as well as fracture face information is present. The etched surface of the membrane can be seen around the edge of the fracture face and intrinsic membrane proteins are easily visualized. Information details revealed by different fracture and etch faces and the standard nomenclature of these faces is given by Branton *et al.* (1975).

5.3.5 Samples for Scanning Microscopy

Sample preparation for scanning electron microscopy is generally much simpler than for transmission microscopy. Inorganic specimens can be examined directly. Biological samples usually have to be treated to withstand the microscope vacuum and this may involve dehydration, but some specimens, like plant leaves, can be studied in the hydrated state.

Two methods of dehydration are popularly used on specimens for scanning microscopy. The first is similar to that used in the thin section technique, namely dehydration through a series of ethanol solutions. The final liquid can be removed by critical-point-drying using $CClF_3$ or CO_2 as intermediate fluids. Critical-point-drying has to be done under pressure but avoids artefacts due to surface tension since at the critical point the distinction between liquid and gas disappears. (It is unfortunate that the pressures required to reach the critical point of water are too great to allow direct dehydration without intermediate fluids.) The specimens have to be stabilized by pretreatment with glutaraldehyde but even so, some shrinkage will occur which can cause structural artefacts. The second dehydration method is to freeze-dry the specimen. Again prefixation is advisable before freezing to keep shrinkage to a minimum (less

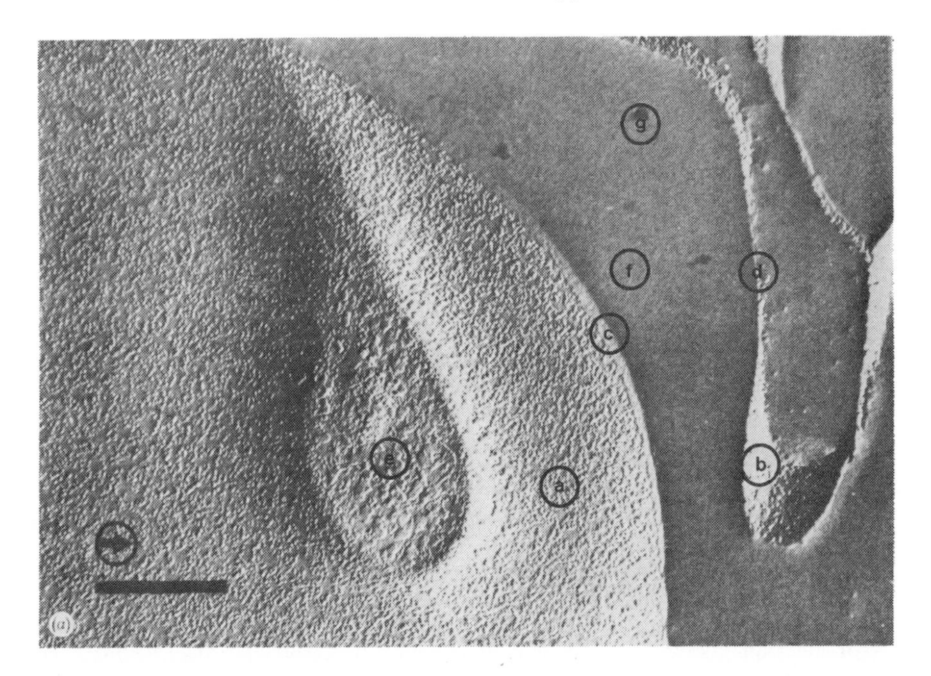

Fig. 13 Erythrocyte membranes examined by the freeze-fracture technique. The purified erythrocyte membranes in 30% glycerol solution were frozen in liquid Freon-22. On the final replica can be seen membrane fracture faces (a and b), a membrane etched surface (c), a cross-sectional view of the membrane (d), ice crystal damage (e), the frozen glycerol eutectic background (f) and contamination of the replica by the sample due to insufficient cleaning after replication (g). Bar = 500 nm. Direction of shadowing indicated by arrow. (Reprinted by permission from Wrigglesworth, 1975)

than 10%). The temperature for vacuum drying is best kept below −40 °C to avoid any recrystallization of ice in the sample.

With the use of a cold stage in the scanning microscope it is possible to study frozen hydrated material, which is especially useful for X-ray microanalysis. Freeze-fracture replicas can also be profitably studied by high resolution (2–3 nm) scanning electron microscopy. Some specificity of surface labelling has been achieved using the antigen–antibody technique with ferritin or colloidal gold particles which can be detected in the scanning mode.

The increase in scanning resolution (to less than 5 nm) has meant that the fine structure of the metal coating, which is often used to render the specimens electron conducting, may obscure fine specimen detail. Vacuum evaporation of metal from a hot wire filament as illustrated in Fig. 10 (but with rotation of the sample) can lead to coating artefacts and even when these are minimized a granular surface structure is generally seen, with crystal sizes between 10 and 15 nm for gold and slightly less for platinum. To avoid this problem the samples may be coated by ion-beam spluttering which has a relatively slow deposition rate but provides a dramatic improvement in coating quality. The technique uses a collimated ion beam, usually argon, to deposit the metal on to the specimen. The argon ions act as nucleating sites for crystallite growth of the coating metal whilst it is in transit in the plasma beam. Collimating the beam and rotating the specimen enables coating to take place with little or no heating effect. Using high vacuum to avoid contamination, this coating method can reduce grain size to less than 2 nm.

Acknowledgment

I thank the staff of the Electron Microscopy Unit at Chelsea College, especially Malcolm Wineberg, for help and advice in preparing this chapter.

References

Agar, A. W., and Keown, S. R. (1978) Measurements from micrographs by optical diffraction. *Proc. Roy. Micros. Soc.* **13**, 147–149

Barnett, W. A., Wise, M. L. H., and Jones, B. L. (1975) Cathodoluminescence of biological molecules, macromolecules and cells. *J. Microsc.* **105**, 299–303

Bjorkman, N., and Hellström, B. (1965) Lead-ammonium acetate, a staining method for electron microscopy free of contamination by carbonate. *Stain Technol.* **40**, 169–171

Black, M. M., and Ansley, H. R. (1966) Histone specificity revealed by ammoniacal silver staining. *J. Histochem. Cytochem.* **14**, 177–181

Branton, D., Bullivant, S., Gilula, N. B., Karnovsky, M. J., Moor, H., Mühlethaler, K., Northcote, D. H., Packer, L., Satir, B., Satir, P., Speth, V., Staehelin, L. A., Steer, R. L., and Weinstein, R. S. (1975) Freeze-etching nomenclature. *Science* **190**, 54–56

Couston, B. E. (1981) Resins: Toxicity, hazards and safe handling. *Proc. R. Microsc. Soc.* **16**, 265–267

Eagles, P. A. M., and Johnson, L. N. (1972) Electron microscopy of phosphorylase b crystals. *J. Mol. Biol.* **64**, 693–695

Frederik, P. M., and Busing, W. M. (1981) Ice crystal damage in frozen thin sections: freezing effects and their restoration. *J. Microsc.* **121**, 191–200.

Geuze, H. J., Slot, J. W., van der Ley, P. A., Scheffer, R. C. T., and Griffith, J. M. (1981) Use of colloidal gold particles in double-labelling immunoelectron microscopy of ultrathin frozen tissue sections. *J. Cell. Biol.* **89**, 653–655

Gordon, M., and Bensch, K. G. (1968) Cytochemical differentiation of the guinea pig sperm flagellum with phosphotungstic acid. *J. Ultrastruct. Res.* **24**, 33–50

Gupta, B. L., and Hall, T. A. (1981) The X-ray microanalysis of frozen hydrated sections in scanning electron microscopy: An evaluation. *Tissue and Cell.* **13**, 623–643

Hall, T. A. (1979) Biological X-ray microanalysis. *J. Microsc.* **117**, 145–163

Haschemeyer, R. H., and Meyers, R. (1972) Negative staining. In: *Principles and Techniques of Electron Microscopy: Biological Applications* (ed. M. A. Hayat), Vol. 2. Van Nostrand Reinhold Company, New York

Hayat, M. A. (ed.) (1973–1977) *Electron Microscopy of Enzymes: Principles and Methods* (6 vol.). Van Nostrand Reinhold Company, New York

Hayat, M. A. (1981) *Principles and Techniques of Electron Microscopy: Biological Applications*. Edward Arnold (Publishers) Ltd, London

Hendy, R. (1971) Electron microscopy of lipofucsin pigment stained by the Schmörl and Fontana technique. *Histochemie* **26**, 311–318

Herbst, R., and Hoder, D. (1978) Cathodoluminescence in biological studies. *Scanning* **1**, 35–41

Holm, R., and Reinfandt, B. (1977) Auger microanalysis in a conventional SEM. *Scanning* **1**, 42–57

Isaacson, M., and Johnson, D. (1975) The microanalysis of light elements using transmitted energy loss electrons. *Ultramicroscopy* **1**, 33–52

Jacob, J., and Budd, G. C. (1975) Application of electron autoradiography to enzyme localization. In: *Electron Microscopy of Enzymes* (ed. M. A. Hayat), Vol. 5, pp. 218–266. Van Nostrand Reinhold Company, New York

Janssen, A. P., and Venables, J. A. (1979) Auger microscopy – an introduction for biologists. In: *SEM 79* (ed. Om Johari), Vol. II, pp. 259–278. SEM Inc., Chicago

Karnovsky, M. J. (1961) Simple methods for 'staining with lead' at high pH in electron microscopy. *J. Biophys. Biochem. Cytol.* **11**, 729–732

Kelley, R. D., Dekker, R. A. F., and Bluemink, J. G. (1973) Ligand-mediated osmium binding: its application in coating biological specimens for SEM. *J. Ultrastruct. Res.* **45**, 254–258

Kuo, J. (1980) A simple method for removing stain precipitates from biological sections for transmission EM. *J. Microsc.* **120**, 221–224

Lewis, P. R., and Knight, D. P. (1977) Staining methods for sectioned material. In: *Practical Methods in Electron Microscopy* (ed. A. M. Gluauert), Vol. 5, Part 1. North-Holland Pub. Co., Amsterdam

Molin, S. O., Nygren, H., and Dolonius, L. (1978) A new method for the study of glutaraldehyde-incurred crosslinking properties in proteins with special reference to the reaction with amino groups. *J. Histochem. Cytochem.* **26**, 412–422

Mowry, R. K. (1958) Improved procedure for the staining of acid polysaccharides by Muller's colloidal (hydrous) ferric oxide and its combination with the Fuelgen and the PAS reactions. *Lab. Invest.* **7**, 566–568

Newbury, D. E. (1976) The utility of specimen current imaging in the SEM. *Scanning Electron Microscopy*. pp. 111–120. IIRI, Chicago

Pscheid, P., Schudt, C., and Plattner, H. (1981) Cryofixation of monolayer cell cultures for freeze-fracturing without chemical pre-treatments. *J. Microsc.* **121**, 149–167

Rash, J. E., and Hudson, C. S. (1979) *Freeze-fracture: Methods, Artifacts and Interpretations*. Raven Press, New York

Ratcliffe, E. (1962) The thermal conductivity of ice: new data on the temperature coefficient. *Phil. Mag.* **7**, 1197–1205

Reed, S. J. B. (1975) *Electron Microprobe Analysis*. Cambridge University Press, Cambridge, UK

Reimer, L. (1978) Scanning electron microscopy – present state and trends. *Scanning* **1**, 3–16

Reynolds, E. S. (1963) The use of lead citrate at high pH as an electron opaque stain in electron microscopy. *J. Cell. Biol.* **17**, 208–214

Robards, A. W., and Crosby, P. (1979) A comprehensive freezing, fracturing and coating system for low temperature scanning electron microscopy. *Scanning Electron Microsc.* **2**, 325–344

Sandritter, W., Riede, U., and Kiefer, G. (1981) A simple method for the use of gallocyanin-chrome alum as an electron stain. *J. Microsc.* **121**, 253–259

Schwartz, L. H., and Cohen, J. B. (1977) *Diffraction from Materials*. Academic Press, New York, San Francisco, London

Skaer, R. J. (1981) A new *en bloc* stain for cell membranes: tannic methylamine tungstate. *J. Microsc.* **123**, 111–113

Soni, S. L., Kalnins, V. I., and Haggis, G. H. (1975) Localisation of caps on mouse B lymphocytes by SEM. *Nature (London)* **255**, 717–719

Statham, P. J. (1981) X-ray microanalysis with Si (Li) detectors. *J. Microsc.* **123**, 1–23

Sternberger, L. A. (1979) *Immunocytochemistry*. John Wiley & Sons, New York, Chichester, Brisbane, Toronto

Sternberger, L. A., Hardy, P. H., Cuculus, J. J., and Meyer, H. G. (1970) The unlabelled antibody enzyme method of immunohistochemistry: Preparation and properties of soluble antigen–antibody (horseradish peroxidase–antihorseradish peroxidase) and its use in identification of spirochetes. *J. Histochem. Cytochem.* **18**, 315–333

Stockenius, W., and Malin, S. C. (1965) Studies on the reaction of osmium tetroxide with lipids and related compounds. *Lab. Invest.* **14**, 458–468

Van Steveninck, M. E., and Van Steveninck, R. F. M. (1981) An X-ray microanalytical examination of precipitation methods for the ultrastructural localization of potassium in plant tissues. II. Tetraphenyl boron. *J. Microsc.* **123**, 51–60

Wrigglesworth, J. M. (1975) Electrostatic interactions at the plasma membrane. *Phil. Trans. R. Soc. Lond. B* **271**, 233–410

Wrigglesworth, J. M., Packer, L., and Branton, D. (1970) Organisation of mitochondrial structure as revealed by freeze-etching. *Biochim. Biophys. Acta* **205**, 125–135

Biochemical Research Techniques
Edited by J. M. Wrigglesworth

6
Monoclonal Antibodies

NORMAN A. STAINES

Immunology Section, Chelsea College, University of London, Manresa Road, London SW3 6LX

6.1 Introduction

Monoclonal antibodies (MCA) are the products of individual clones of lymphocytes. Used singly, free of contamination by antibodies made by other lymphocytes, they are powerful immunologically specific reagents and analytic probes.

This chapter describes the technology that has been developed to prepare MCA of preselected specificity. They are of enormous potential value in genetics, biology, biochemistry, serology and in disease diagnosis and therapy. This technology is based on a procedure which was described by Köhler and Milstein in 1975 and has in only a few years revolutionized immunology: it has increased dramatically the resolving power of serological technique. The chapter is concerned with how MCA can be used and, more extensively, with the problems involved in preparing them. References are given on points of particular importance so that interested readers may pursue them in order to decide if the technology could be applied to their scientific problems.

6.1.1 The Clonal Complexity of the Immune Response

Even relatively simple antigens usually express several different antigenic determinants (epitopes) which induce the formation of a complementary but diverse set of specific antibodies, each of which is different in that it reacts with only one epitope on the antigen. Further, a single epitope itself may induce the formation of antibodies of different affinity and immunoglobulin (Ig) class. Each individual antibody is monoclonal – the product of one expanded lymphocyte clone. Antisera contain oligoclonal mixtures of antibodies.

The clonality of the antibody response is even more diverse when the immunizing antigen is impure. The contaminants induce antibody formation often more efficiently than the antigen in question. This is because of variations in the intrinsic immunogenicity of the components of the immunizing material and because of antigenic competition. In essence, there is no reliable way of predicting the specificity of antibodies that can be induced by a particular antigen: the production of specific antisera is largely an empirical exercise. Unwanted antibodies can sometimes be removed by absorption, but this may produce complexes of antigen and antibody which modify the properties of the antiserum. Even if irrelevant antibodies are not found, the suspicion often remains that they are merely undetectable in the assay used. This is a considerable problem where the specificity of an antiserum is established in one system (e.g. immunodiffusion) which by necessity is different from that (e.g. immunohistochemistry) in which it is ultimately used experimentally.

6.1.2 Monoclonal Antibody (MCA) Reagents

These points illustrate some of the difficulties in producing and using oligoclonal antisera. It is true that we are almost always in ignorance of which antibodies within a particular antiserum are responsible for the effects it has. The antibodies vary in not only specificity and affinity but also Ig class which controls their biological effector functions, unrelated to their specificity. Awareness of problems such as these led many scientists to believe that if antibodies could be produced in monoclonal form then a new dimension of accuracy and precision could be brought to immunochemistry. Because there are no obvious preparative methods of separating MCA from an antiserum, a radically different technology based on separating the cells that make the antibodies is required. Fortunately, this is available and has been used successfully to make MCA specific for all classes and types of antigens.

Each MCA has a precisely definable specificity and affinity, is of one Ig class, and can be prepared in large amounts. The technology involved enables the production of pure antibodies from animals immunized with impure antigens. It has been most widely used to produce MCA from mice and, less extensively, from rats. It is now possible to prepare human MCA and although at the time of writing this is not widespread it can only be a matter of time before it is: the motivation is overwhelming to produce human MCA for therapeutic use. It must be said, however, that for many other uses, including clinical diagnostic tests, human MCA may not necessarily offer specific advantages because rodents are quite able to produce antibodies that identify micropolymorphisms of heterologous proteins. Further, the ease with which rodent immune responses can be manipulated must always make animals more useful than man in many applications of this technology.

6.1.3 Resources Required for Producing MCA

The parts of the technology involve:

(1) hybridizing antibody-secreting lymphocytes with myeloma cells to make somatic cell hybrids which secrete specific antibody;

(2) separating different lymphocyte hybrid cells from each other by cloning; and
(3) propagating them as cell lines (hybridomas) each making one MCA.

Production of hybridomas and MCA involves a considerable investment of resources. There are two facts of paramount importance to consider before making this investment. First, it takes several months to isolate stable cloned cell lines making MCA in even the simplest system. Second, the amount of long-term tissue culture involved is such that the culture facility must be maintained reliably and available continuously. There is no doubt that the most efficient way to produce MCA is to dedicate a facility to doing nothing else, and to employ trained culturists in it. In this way the needs of several research groups might be met by one culture laboratory. It is the case, however, that many scientists are obliged, or wish, to prepare MCA themselves and to do this the following basic equipment is essential:

(1) a work station for handling cell cultures; a vertical laminar flow sterile air cabinet is ideal.
(2) an incubator that can be maintained at 37 °C with a humidified (saturated) atmosphere of 5% CO_2 in air. A water jacketed incubator with an air circulating fan is preferable.
(3) a liquid nitrogen storage and freezing facility for cell lines.
(4) the usual supporting facilities for tissue culture work including sterilizing oven, refrigerator, autoclave and deep freeze.
(5) emergency back-up incubator, culture cabinet and deep freeze. The importance of these cannot be over-estimated if equipment failure is not to result in catastrophic losses of unique cell lines.

It is possible, with attention to detail, to maximize the chances of producing MCA of the required specificity and class. The time required to do this and the technical problems involved may be extensive. Training and facilities are equally important and anyone entering this area should seek the help and advice of colleagues already involved in it. The intending producer who cannot commit resources to the activity might propose to collaborate with an established hybridoma laboratory. Conversely, it is an argument of some merit that any sizeable immunology laboratory should consider setting up their own service facility for making MCA because the hybridoma technology has so many obvious and important applications.

6.2 Methodology of Producing MCA

6.2.1 General Principles

Since MCA are produced by single cells and their clonal progeny, then the physical separation of lymphocytes is obviously a central step in preparing

MCA. It is technically simple to separate cells physically and this can be done in several ways. Lymphocytes present other problems, however, in that they do not secrete antibody indefinitely and will not survive *in vitro* beyond a few days or at best a few weeks. Thus, the immortalization of the lymphocytes in a way in which they continue to secrete specific antibodies is the essential prerequisite to cell cloning and producing MCA. It must be said at this stage that there are many methods for preparing MCA that differ only in detail: in these principles, as illustrated in Fig. 1, they are the same:

(1) Immunize mice with antigen against which MCA are required.
(2) Prepare an immune lymphocyte cell suspension and mix with suitable myeloma or plasmacytoma cells.
(3) Expose the cell mixture to a cell-fusion promoting agent.
(4) Divide the fused cell mixture into many small samples and maintain in selective tissue culture conditions that permit the survival of hybridized cells.
(5) Identify cultures containing MCA of the required specificity.
(6) Recover cells from these cultures and isolate, by repeated cycles of cloning, stable cell lines each secreting one MCA.
(7) Freeze seed stocks of each line for future use.
(8) Grow cloned cell lines *in vitro* or *in vivo* as hybridomas and recover the secreted MCA.

The ways in which MCA can be prepared by lymphocyte hybridization and cloning have been described by many authors. Useful detailed and general information can be found in articles by Oi and Herzenberg (1980), Goding (1980) and Galfré and Milstein (1981) and in books edited by Melchers *et al.* (1978), Lefkovits and Pernis (1979), Kennett *et al.* (1980), and Hämmerling *et al.* (1981).

6.2.2 Immunization and Generation of Immune Lymphocytes

6.2.2.1 The Purpose of Immunization

Specific MCA can only be derived from animals making the antibodies in the first place. The number of hybridomas that can be prepared is related in a direct but complex way to the frequency of antibody-producing cells in the lymphocyte source used for hybridization. For these reasons it is important that cells are taken from animals immunized appropriately.

For many antigens it may already be obvious how mice can be immunized optimally to produce specific serum antibody but there are unfortunately very few general rules applicable to immunization. Thus, to produce specific MCA the best procedure for each antigen needs to be established individually.

The aim of immunization is to expand selectively the clone or clones of

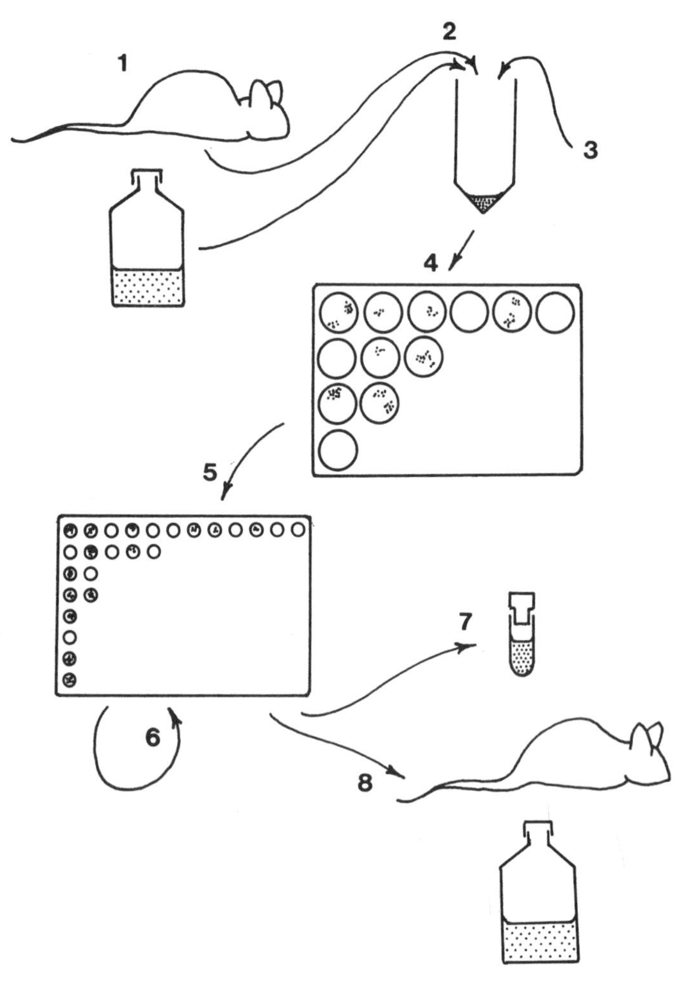

Fig. 1 Preparation of monoclonal antibodies. (1) Immunize mouse with antigen. (2) Prepare a suspension of lymphocytes from the immune mouse and mix these with suitable myeloma cells. (3) Expose the mixed spleen–myeloma cell suspension to a cell-fusion promoting agent. (4) Wash the cells, resuspend in HAT tissue culture medium, add feeder cells and distribute into many 1 ml cultures in 24 well multiplates. (5) When cell growth is visible, screen for antibody production. Take cells from positive cultures and clone at limiting dilution in 96 well multiplates. (6) Screen clones for production of antibody, and repeatedly reclone positive cultures to produce stable monoclonal cell lines, each secreting one MCA. (7) Freeze seed stocks of each line for future use. (8) Grow the cloned cell lines in ascites form in mice or in culture *in vitro* and harvest secreted antibody

lymphocytes making antibodies of the required specificity. For antigens that are impure, poor immunogens, have many epitopes, or are toxic, there are a number of different tactics that may be used to increase the probability of producing the right hybridomas.

6.2.2.2 Immunization Schedules

It might be expected that cells from hyperimmunized animals would ultimately yield more specific hybridomas than cells from animals less extensively immunized. There is, however, enough general experience to indicate that this is not always the case: large numbers of hybridomas can be derived from animals after only one, or a few injections of antigen spaced closely together (Hämmerling *et al.*, 1981). Furthermore, IgG MCA and IgM MCA can be derived from animals that have been immunized respectively once or several times. As a first approximation it is reasonable to immunize in a way which produces a high titre of serum antibody. For most soluble, particulate or cellular antigens in mice this means repeated injections 1–2 weeks apart. The rise in serum antibodies is measured and when at a maximum, the animals then used. Whatever schedule is used for immunization, the last injection should be given 3–4 days before harvesting immune lymphocytes. This holds for most systems that have been studied, immune lymphocytes taken earlier or later yield lower numbers of hybrid clones secreting specific antibody. It appears that the cell that best forms antibody-secreting hybrids is a preplasma cell (plasmablast) in which the antibody synthesizing machinery is activated but which has not yet undergone the irreversible cellular changes that lead to the loss of function in mature plasma cells.

The last injection is usually given intravenously in order to deliver the antigen to the spleen cells. However, if strong local responses occur in regional lymph nodes, their cells can be used satisfactorily for fusion. The peritoneal cavity also contains lymphocytes that can be hybridized (Pages and Bussard, 1978). The relevance of circulating antibody levels to the number and specificity of antibody forming cells should not be overestimated. While they are a guide to the state of immune activation there are limitations in titration systems which restrict the isotype, specificity or affinity of the antibodies detectable. Some antibodies may persist after the cells making them have ceased to function. Conversely, many cells may be making antibodies undetectable in the serum. Within an inbred mouse strain it is common to find quantitative variations between individuals in their immune responses. Thus a judicious selection of individual immune mice on the basis of the titre and specificity of their serum antibodies may be useful. Alternatively, several individual fusions can be done simultaneously with cells from different animals.

The number of antibody-producing cells in a population of lymphocytes may be expanded by 'parking' immune cells in an irradiated syngeneic animal for a few days before final antigenic stimulation and cell hybridization (Kenny *et al.*, 1981).

6.2.2.3 *Immunological Adjuvants and Antigen Modification*

It may be necessary to use adjuvants. Freund's Complete Adjuvant (FCA) for the first injection and Freund's Incomplete Adjuvant (FIA) for the second and subsequent injections are widely used and are best injected subcutaneously or intradermally divided into several sites. Killed *Bordatella pertussis* organisms have a strong adjuvant effect for some antigens in mice. Alum-precipitated antigen may also have increased immunogenicity. It must be remembered that some adjuvants are themselves immunogenic and that they all may induce a spectrum of antibodies against the specific antigen that is different from that induced by antigen alone. For a general review on adjuvants see Anon (1976) and Whitehouse (1977).

Soluble antigens with low intrinsic immunogenicity may be linked chemically to a carrier molecule. The carrier chosen should be a T-cell dependent antigen to which the mouse itself responds well.

6.2.2.4 *Selection of Animal Strain for Immunization*

Strains of inbred mice (and rats) may vary in their responsiveness to specific antigens and it is therefore useful to screen several for their response to the antigen in question to select the one which responds best.

F_1 hybrid progeny prepared from different strains often respond immunologically better than homozygous parental animals. It has been suggested that the High Responder mouse strain of Biozzi is a good source of cells for hybridoma production (Boumsell and Bernard, 1980). The murine myeloma cells commonly used to prepare hybridomas are derived from BALB/c mice. Accordingly, if the cell fusion employs immune lymphocytes from F_1 hybrid mice or non-inbred animals (such as Biozzi strains) then particular, and possibly complex, histocompatibility requirements will have to be satisfied if the resultant hybridomas are to be grown as tumours in mice.

6.2.2.5 *Antigen Purification*

The hybridoma technique makes possible the production of pure antibodies against impure antigens. As a general rule, the purer the antigen the greater will be the proportion of derived hybridomas which secrete specific MCA. However, contaminants below the level of chemical detection can be powerful immunogens and thus the effort of purifying antigens extensively, assuming a procedure for doing so is available, must be balanced against both this and the efficiency of cell cloning later as a purification method itself to produce individual MCA. Frequently, however, purification procedures are not available and impure antigens are used by necessity to immunize. In these cases a high proportion of the hybridomas secrete irrelevant antibodies. If the specific MCA can be identified easily then this presents no particular problems to selecting the cultures to be cloned. On the other hand, if this cannot be done, then cell lines must be cloned before the specificity of their MCA against the immunizing complex can be established. In this way, MCA against contaminants will almost certainly be made and this may have a serendipitous value in providing unexpectedly useful specificities. Milstein and Lennox (1980) have suggested how these MCA can be used to practical advantage to make immunoadsorbent columns with which impure antigen preparations are depleted of contaminants and used for another cycle of MCA production. If only MCA against contaminants are again recovered they can be employed in a similar manner, and, although laborious, this type of cascade purification scheme must finally yield the desired specific MCA.

6.2.2.6 Selective Suppression of Antibody Production Against Irrelevant Antigenic Determinants

If antigens are only available in an impure form for both immunization and later for screening culture supernatants, the job of producing MCA and determining their specificity becomes very difficult. In attempts to depend upon something other than chance to give success, consideration should be given to ways of minimizing the production of irrelevant antibodies, by, for example, the induction of immunological tolerance to contaminating antigens. This could be particularly useful in preparing MCA against cell surface antigens where a different cell type could be used as a tolerogen, or against Ig idiotypes or isotypes where normal Ig could be used as the tolerogen.

Passively administered antibodies can specifically suppress immune responses. Antibodies reactive with some of the antigens in the immunizing mixture have been used to restrict the clonality of the immune response to derive MCA specific for neuroblastoma antigens (Kennett and Gilbert, 1979). However, the mechanism of this type of phenomenon (passive enhancement) is complex and antibody against one epitope is likely to prevent formation of antibodies against other epitopes on the same molecule (or cell) so this procedure might be most suitable for restricting responses against contaminants physically separate from the specific antigen.

6.2.2.7 Polyclonal Activation of Lymphocytes and Antibody Libraries

Normal lymphocytes exposed to polyclonal activators such as bacterial lipopolysaccharide or pokeweed mitogen secrete antibodies of many different specificities. Such lymphocytes have been used to derive hybridomas secreting

MCA against a variety of antigens without the cells being first exposed to the specific antigens. Melchers *et al.* (1978) list a number of hybridomas so derived.

In theory, this could be a way of producing a library of MCA representative of the antibody repertoire of an individual mouse. Large numbers of Ig-secreting hybridomas derived from lipopolysaccharide activated spleen cells were examined by Andersson and Melchers (1978) who found that about 35% produced MCA that bound either to heterologous erythrocytes or phenyl derivative haptens. These authors suggest that this strategy could be used to produce MCA specific for antigens that are too toxic for immunization. It is not clear that polyclonal activation of lymphocytes necessarily induces the same spectrum of antibodies, particularly with respect to affinity, as that made in response to specific antigenic stimulation.

In producing MCA against a specific antigen, the number of hybridomas sought will obviously depend upon the uses to which the MCA will be put and it is necessary to consider this issue before embarking upon their production. In many cases a single MCA, or at least a small number of them might be adequate but in others it will be necessary to produce an extensive panel of specific MCA of different affinity and class. The effort required to produce them is directly proportional to their number. It should be remembered that even simple antigens may activate many different B-lymphocytes. There are, in consequence, potentially many different specific MCA for a given antigen and in some heterologous protein antigenic systems it is uncommon to find a particular antibody repeated in a panel of several dozen specific MCA. This extensive spectrum of antibody specificity is worthy of study in its own right and must be considered in any plan to prepare specific MCA either singly or as libraries.

6.2.3 Myeloma Cells for Hybridization

The myeloma cells for hybridization with immune lymphocytes must have the basic property of not inhibiting the reading of the genome acquired from the immune lymphocyte and the secretion of antibody molecules in the daughter hybrid cells. Cotton and Milstein (1973) demonstrated that hybrids of two different myeloma cells produced the H and L chains of both parental cells and thus had complemented Ig genes. This showed that hybrid lymphocytes are able to synthesize Ig and in the early successful immune lymphocyte × myeloma fusions the hybrid cells produced both specific antibody H and L chains and the γ_1- and κ-chains of the X63-Ag8 parental myeloma (Köhler and Milstein, 1975, 1976) or the γ_{2b}- and κ-chains of the X45.6TG1.7 parental myeloma (Margulies *et al.*, 1976). The Ig molecules secreted by the cloned lines were heterogeneous assemblies of the four different chain types. Such antibodies, although strictly speaking monoclonal, are not homogeneous. The rates of mutation and loss of expression of Ig genes can be relatively high in myeloma cells (Coffino and Scharff, 1971, Cotton *et al.*, 1973) and it is possible to select chain-loss variants of hybrid lymphocytes which secrete Ig molecules composed of only the antibody H and L chains (Köhler and Milstein, 1976).

It is obvious, however, that it would be better to use myeloma cells that do not themselves secrete Ig chains but would still permit the expression of Ig genes in the complemented genome of hybrid cells derived from them. Accordingly, several such lines have been derived and are now available to make MCA (Table 1). It is worth noting that the P3/NS1/1-Ag4-1 murine line (Köhler and Milstein, 1976) which is widely used does not itself secrete myeloma Ig chains but does contain intracellular κ-chains. Hybrid cells derived from it may secrete these myeloma κ-chains incorporated at random into the MCA molecules.

All the cell lines have been adapted for continuous culture *in vitro*. Because of the need to use selective media at a later stage (to permit the growth of hybrid cells and prevent unfused myeloma cells from growing) they are mutational variants of the parent myelomas: they lack expression of the enzyme hypoxanthine guanine phosphoribosyl transferase (HGPRT). The conditions for maintaining the cell lines are described in references given in Table 1. The rodent lines can be adapted to grow in either RPMI1640 Tissue Culture (TC) Medium or Dulbecco's Modified Eagle's TC Medium. Both media are equally suitable and are described in detail by Paul (1975). Seed stocks of myeloma lines are cryopreserved, and when required are resuscitated and maintained in logarithmic phase culture. It is advisable to use cloned stocks which are exposed periodically to the original selective agent (e.g. 8-azaguanine) to prevent $HGPRT^+$ revertant mutants becoming established.

The originators give some data on the performance of the lines in producing hybridomas (Table 1 references) but there have been no comprehensive analyses of the relative efficiencies of all the cell lines in this. Many laboratories prefer particular lines but it has to be acknowledged that some technical procedures do not transpose well in every detail between laboratories. It is sensible therefore, when establishing hybridoma technology, to obtain several different myeloma lines or sublines and select that which gives the best results for local conditions. Once prepared, hybridomas of most myeloma origins appear to have similar stabilities. It is to be expected that as the processes involved in lymphocyte hybridization become better understood then new cell lines with better characteristics will be isolated.

6.2.4 Fusion of Immune Lymphocytes and Myeloma Cells

The fusion of somatic cells to form binucleate and multinucleate forms is a natural, if uncommon, event. If two nuclei in a dikaryon divide synchronously they may fuse to produce a tetraploid hybrid cell which expresses functions of both the parental cells. Exogenous agents are used to increase the probability of membrane fusion which in turn increases the frequency of formation of

Table 1 Myeloma cell lines suitable for producing hybridomas

Species	Myeloma derived line	Parental line	Ig chains produced[a]	Reference
Mouse	P3/X63-Ag8	P3K(MOPC 21)[b]	$\gamma_1\kappa$	Köhler and Milstein (1975)
	P3/NS1/1-Ag4-1	P3/X63-Ag8	$-\kappa$[c]	Köhler and Milstein (1976)
	Sp2/0-Ag14	X63xBALB/c hybridoma	None	Shulman *et al.* (1978)
	X63-Ag8.653	P3/X63-Ag8	None	Kearney *et al.* (1979)
	FO	Sp2/O-Ag14	None	Fazekas de St Groth and Scheidegger (1980)
	MPC-11-X45-6TG1.7		$\gamma_{2b}\kappa$	Margulies *et al.* (1976)
	S194/5XXO.BU.1		None	Trowbridge (1978)
Rat	210-RCY3-Ag1	Lou rat myeloma	κ	Galfré *et al.* (1979)
Man	U-266AR$_1$	U-266 myeloma	$\varepsilon\,\kappa$	Olsson and Kaplan (1980)
	GM-1500-6TG-A12	GM-1500	$\gamma_2\,\kappa$	Croce *et al.* (1980)

[a] Ig heavy chains are of $\alpha, \gamma, \delta, \varepsilon$, or μ and light chains of κ or λ isotypes. See for example McConnell *et al.* (1981) for general description of these.
[b] Horibata and Harris (1970)
[c] Only secreted after hybridization

viable hybrid cells. Somatic cell hybridization is described in detail by Ephrussi (1972) and Ringertz and Savage (1976).

A general feature of cell hybridization that is of central importance to MCA technology is that the genetic relationship between the cell types fused together governs the functioning and stability of the complemented genome of the hybrid cell. Hence, myeloma cells produce stable hybrids preferentially with B-lymphocytes rather than T-lymphocytes and mouse myeloma × mouse B-lymphocyte hybrid cells tend to be more stable than mouse myeloma × human B-lymphocyte hybrids. Mouse × human hybrids can be made but their stability is low and the amounts of MCA they secrete are less than those secreted by intraspecific hybrids (Schwaber, 1975: Croce *et al.*, 1981). For these reasons, and because there are few human myeloma lines suitable for lymphocyte hybridization (Croce *et al.*, 1980; Olsson and Kaplan, 1980), MCA production has been limited largely to the mouse and rat systems.

Some chromosomes may segregate unequally in replicating hybrid cells and this leads to the loss of functions controlled by them. This is probably the main reason for the loss of Ig production in lymphocyte hybridomas in general. The underlying mechanisms of chromosome loss are not understood, but, for example, as Schröder *et al.* (1981) discuss, the segregation or retention of human chromosomes in heterospecific hybrids may be a non-random event. If this could be manipulated, then stable interspecies hybridomas might be produced more easily and this would avoid the need to find myelomas suitable for fusion from every species.

6.2.4.1 Technique of Fusion

Immune lymphocytes prepared as a monodisperse suspension are mixed with myeloma cells, the mixture is washed, concentrated and exposed under carefully controlled conditions to a fusing agent. Sendai virus was used originally in experiments on immune lymphocyte hybridization (Köhler and Milstein, 1976) but this is not the ideal fusing agent for murine lymphocytes and it is common practice now to use polyethylene glycol (PEG) employed first for this purpose by Pontecorvo (1975) and for lymphocytes in particular by Galfré *et al.* (1977). PEG has the advantage of being inexpensive and easily obtainable. Detailed descriptions of techniques for using PEG to promote cell fusion are given by Davidson and Gerald (1976). The techniques most favoured for lymphocyte fusion involve adding PEG in solution to lymphocyte-myeloma cells either in concentrated suspension in a tube (Galfré *et al.*, 1977; Kennett *et al.*, 1980; Oi and Herzenberg, 1980) or deposited on polycarbonate filter discs (Buttin *et al.*, 1978)

Careful attention must be paid to the conditions used to expose cells to PEG for it can be highly toxic. It is necessary also to screen batches of PEG for toxicity to lymphocytes under controlled conditions. PEG preparations with mean molecular weights between 1000 and 6000 are suitable for cell fusion; outside of this range they have unacceptably high toxicity. The addition of dimethylsulphoxide (DMSO) may minimize the cell toxicity of PEG (Schneiderman *et al.*, 1979, Hämmerling *et al.*, 1981). The optimum concentration of PEG is normally between 30 and 50% in a serum-free medium, such as phosphate buffered saline (PBS) containing Ca^{2+} and Mg^{2+}. The time for which cells are exposed to PEG at this concentration is usually short (1 minute) to minimize the membrane damaging effects of the fusing agent. The cells are then gently resuspended, diluted and washed free of PEG. This fusion and washing process should be conducted at 37 °C. This is discussed fully by Oi and Herzenberg (1980), Fazekas de St. Groth and Scheidegger (1980), Kennett *et al.* (1980), and Goding (1980).

It is not necessary to purify B-lymphocytes before fusion because myeloma cells selectively fuse with these cells. Thus, it is conventional in many of the techniques to use suspensions of whole spleen cells for fusion: also it is not essential to remove erythrocytes by flash lysis which anyway could damage the lymphocytes in the process.

The ratio of spleen:myeloma cells in the fusion mixture influences the number of hybrid cells produced although not in a simple way. It is a general feature of cell hybridization that more hybrids are produced from unequal mixtures of parental cells and this holds also for lymphocyte fusion. Success has been recorded with spleen:myeloma ratios ranging from 10:1 to 1:2. In all cases the myeloma cells are actually in excess of those lymphocytes (preplasma cells) with which they have the potential to form stable hybrids. It would appear to be sensible to determine by experiment the optimal cell ratio for each particular immune lymphocyte fusion system, but in the absence of such information then a spleen:myeloma ratio of 5:1 or less should be used.

6.2.4.2 Modification of the Fusion Technique

There are many ways of fusing somatic cells and some particular proposals for improving the rate of fusion of lymphocytes with myeloma cells deserve to be examined further. Kranz *et al.* (1980) have indicated that the production of hybridomas secreting antibodies reactive with fluorescyl residues was improved by fluoresceinating the Sp2/0-Ag14 myeloma cells. The use of specific antigen bridging in this way between myeloma and immune lymphocytes might have a more general applicability. Stähli *et al.* (1980) have emphasized the importance of obtaining a high frequency of specific antibody forming cells in the fusion mixture and the use of rosetting techniques to enrich immune lymphocyte populations immediately before fusion (Kenny *et al.*, 1981) may be generally useful for this purpose.

6.2.4.3 Alternatives to Hybridization as a Means of Immortalizing Immune Cell Function

It was known for some years (see review by Sinkovics, 1981) before Köhler and Milstein (1975) described the successful and purposeful production of specific MCA by hybridomas that hybrid lymphocytes could secrete immunoglobulins with antibody activity. While attempts to hybridize mouse lymphocytes were being made, those laboratories interested in immortalizing human lymphocytes were taking a different approach to the problem. This had its origins in the facts that there are few human myeloma cell lines adapted for tissue culture and that viral transformation was known to be involved in the development of human B-lymphoblastoid lines.

Although some human myelomas are now available for hybridization, an alternative technology has been developed for human lymphocytes that depends upon the ability of Epstein–Barr virus (EBV) to infect specific antibody-producing lymphocytes and transform them into continuous cell lines that can be cloned by the usual techniques described later (Steinitz *et al.*, 1977, 1979).

This technique is not widely used, some authorities believing that hybridization is technically easier than virus transformation and yields more stable cell lines secreting larger amounts of antibody. The potential of using a combination of viral transformation and hybridization for producing human cell lines is described by Steinitz (1981).

6.2.5 Positive Selection of Hybridized Lymphocytes in Culture

The fusion cell mixture is washed and resuspended in culture medium and aliquotted in small cultures maintained usually in multiwell tissue culture trays (see below). The early stages of culture are designed to promote the growth of hybrid lymphocytes at the expense of other cells. Non-hybridized spleen cells have only a limited growth potential and do not therefore present the same competitive threat as unfused myeloma cells to the growth of hybrid cells. The fusion conditions are such that the vast majority of the myeloma cells are not fused to spleen lymphocytes and these need to be destroyed. This is accomplished by culturing the cells in the presence of aminopterin or amethopterin, drugs which inhibit folic acid metabolism and which therefore kill the $HGPRT^-$ myeloma cells (Section 6.2.3). Complemented hybrid cells express the HGPRT gene acquired from the immune lymphocyte parent and will survive if given exogenous hypoxanthine and thymidine in the medium. This is the so-called HAT selective medium of Szybalski *et al.* (1962) used here as described by Littlefield (1964) to select positively for normal × malignant cell hybrids.

Other drug-based culture systems have been described for selecting complemented hybrids in culture (see Ringertz and Savage, 1976 for discussion) but the only one of significance here employs ouabain. This can be used in isolating interspecific hybrids: human cells are three to four orders of magnitude more sensitive to its toxic effects than are mouse cells.

The frequency of formation of viable hybrid cells is low (see below). Most cells die at an early stage and this can be indirectly damaging to the surviving cells. This may be minimized by changing the HAT culture medium frequently as described originally by Köhler and Milstein (1975), but an alternative is to leave the cultures undisturbed for up to one week and to supplement the cultures directly after fusion and before plating out with a normal feeder cell population. Many different cell types have been recommended for this and those commonly used include spleen cells, thymus cells, peritoneal exudate cells, erythrocytes or irradiated fibroblasts. All effectively increase the survival of hybrid cells, and the phagocytic macrophages in peritoneal or spleen cells have an added advantage in that they cleanse cultures of dead cells.

The initial distribution of fused cells into many small cultures is a first step towards separating clones of hybrid cells from each other. After a few days, small clonal growth foci are visible in the multiwell cultures and if the fused cell mixture has been diluted correctly there will be only a few clones growing in each culture. Not all clones growing in HAT medium secrete specific MCA. Some are $HGPRT^{+}$ revertant myeloma cells (arising at a rate of 10^{-5} to 10^{-6}) but most are hybrid cells, the majority of which secrete monoclonal Igs of unknown specificity rather than specific antibody against the immunizing antigen.

Two types of flat-bottomed multiwell culture system are used: 1–2 ml cultures in 24 well TC plates and 0.2 ml cultures in 96 well TC plates. The number of clones that can be derived from a single fusion depends upon many factors and varies in a complex way with cell concentration, culture volume and the feeder cell type. As a guide, 10^{8} spleen lymphocytes fused with 2.10^{7} NS-1 myeloma cells and plated (with an equivalent number of normal spleen feeder cells) into 480 2-ml cultures may produce up to 50 clones per culture within 2 weeks.

Hybrid lymphocytes are cultured under standard TC conditions. The medium used most frequently is either RPMI 1640 or DMEM supplemented with fetal calf serum (FCS) up to a level of 20%. Because of the cost and periodic scarcity of FCS there can be advantages in using other serum sources such as horse, pig or newborn calf. In each case, the serum must be free of virus and *Mycoplasma* contamination and should be screened for toxicity against at least the parent myeloma and also, if possible, some hybridoma lines. The most sensitive measure of serum toxicity is its effect upon cell cloning efficiency (Oi and Herzenberg, 1980 describe how this can be assessed) although a reasonable approximation may be achieved by determining its effect upon logarithmic growth rates of cloned cells. Additionally, all TC sera should be checked for potential interference in the assay systems to be used in screening for antibody production by hybrid cells. This is particularly important with TC sera, other than FCS, which contain Ig.

Hybridoma cells are not tolerant of alkaline culture conditions and should therefore be maintained in an atmosphere containing at least 5% CO_2. Culture media freshly supplemented with glutamine should be used. Penicillin and streptomycin are usually employed and antifungal agents are recommended by some authors.

6.2.6 *Isolation of Hybrid Cell Lines Secreting Specific MCA*

Cells are maintained for about 2 weeks after fusion in the basic serum-supplemented HAT medium. After this time they can be adapted for growth in normal medium by a transitional phase in HT medium (Davidson and Ephrussi, 1970) which (presumably) allows the induction of enzymes involved in folic acid metabolism.

When clonal growth is first visible the cultures are screened for the presence of specific antibody in the supernatant and again at later stages according to the rate of growth. The tactics of screening procedures are discussed in the next section.

Cultures containing specific antibody will almost certainly include many cells that do not secrete the antibody. In order (1) to prevent the non-productive cells overgrowing the useful hybrids and (2) to separate the different useful hybrids from each other, it is essential to isolate them by cloning as quickly as possible. This is done by taking the cells from each individual antibody-containing culture, diluting and reculturing them at low concentration either in a gel phase or in a fluid phase TC system in 96 well plates. In either case the principle is the same: dilute the cells and distribute them in a culture system in such a way that at a limiting dilution they will grow as discrete clones that can be harvested free of contaminating cells. Both systems are widely used and descriptions of gel phase systems are given by Pluznik and Sachs (1965), Cotton *et al.* (1973), Paul (1975) and Goding (1980) and of fluid phase systems by Oi and Herzenberg (1980). Galfré *et al.* (1980) discuss clonal competition between, and stability of, hybridomas in culture.

Clones of cells growing in semisolid media (usually containing agarose) can be plucked out with a fine pipette and maintained in small cultures for 24–48 hours and the supernatants examined for antibody activity. Clones in gel producing antibody can also be identified by direct or indirect plaque overlay methods (Köhler and Milstein, 1975), immune-precipitate formation in the gel (Cook and Scharff, 1977) or by replica immunoadsorption plating (Sharon *et al.*, 1979).

It is technically simpler to clone cells in fluid phase media. In this case it is only necessary to sample culture supernatants for antibody activity. If large numbers of cloning plates have to be examined, consideration should be given to using semi-automated multiple sampling methods such as the transfer plate system described by Schneider and Eisenbarth (1979).

Cloning can be more efficient if the antibody-forming hybrid cells can be enriched beforehand. Parks *et al.* (1979) have described how a fluorescence activated cell sorter can be employed to purify antibody-producing cells by exploiting their ability to bind fluorescent plastic microspheres coated with specific antigen. Although the equipment for this is not widely available, the technique clearly has a tremendous potential for simplifying the rapid isolation of cloned cell lines.

It is necessary to clone a hybridoma several times to ensure that it is stable, monoclonal and secretes only one antibody. This will only be achieved if (1) cells are cloned from an average concentration of less than one cell per microculture (or are extensively diluted in semisolid cultures), and (2) all daughter clones secrete the same specific antibody. The mathematical treatment of limiting dilution culture systems to adjudge monoclonality is described by Oi and Herzenberg (1980) and discussed at length by Lefkovits and Waldmann (1979).

The main problem encountered in cloning hybridomas is that some cell lines have a critical limiting density for growth that is too high to permit one cell to proliferate in, say, 0.1 or 0.2 ml culture medium. In other words, their cloning efficiency is low and for any cells to grow at all in such a volume, more than one may need to be present. This restriction may prevent cloning of some B-lymphocyte cell lines and is notably a problem with T-lymphocyte lines. Cloning efficiency is improved considerably if feeder cells are added in the form of normal peritoneal exudate or spleen cells at a concentration of about 10^4 per 0.1 ml culture. Insulin added at 1 i.u./litre to the culture medium can also improve cloning efficiency.

The repeated cloning of many cell lines involves a considerable amount of work and it is desirable to adopt procedures that will minimize this. If possible, therefore, only cultures derived from cells plated initially at the lowest concentration should be recloned and those daughter clones chosen which show the highest growth rate and greatest production of antibody. It is usually adequate at each cloning step to retain only three antibody-producing cultures: reclone one and preserve the others (by freezing) for recovery in the event that the first cannot be recloned successfully. The time required to clone individual hybridomas varies greatly. Some never achieve a high cloning efficiency and in general the doubling times of hybridomas in logarithmic growth range from less than 10 to more than 40 hours.

When a cell line satisfies the criteria for monoclonality it is expanded in culture and a seed stock frozen down for future use because, apart from losses through microbial contamination, many lines prove to be unstable through either mutational changes or chromosome loss. Most murine hybridomas ultimately stabilize with a subtetraploid number of chromosomes but as there does not appear to be any way of exerting a selective pressure on cells to retain chromosomes carrying Ig genes these may be lost before the line stabilizes.

6.2.7 Screening Procedures for Identifying Hybrid Cells Secreting Specific Antibodies

Parts of the technique of producing MCA discussed previously involve the logical application of tissue culture methods which in themselves are well defined and reliable. On the other hand, there is no one general-purpose technique for identifying MCA secreted by hybrid cells. The assay method used will depend upon the antibody–antigen system under consideration and upon the scale of the operation. The efficiency of the screening procedure is central to the preparation of MCA and can be the limiting factor in its success. The choice of an appropriate assay procedure will need to account for the following:

(1) Thousands of individual culture samples may have to be examined at one time.
(2) Clonal competition can lead to rapid elimination of slower growing cells (which may be the antibody-producing cells because of their greater nutritional requirements).
(3) Individual MCA are each of one isotype, specificity and affinity.

It follows, therefore, that the assay should ideally be simple, cheap, quick to perform and be able to handle large numbers of culture supernatant samples and to identify MCA of the appropriate specificity, affinity and class. The assay should be appropriate to the use to which the MCA ultimately will be put. For example, radioimmunoassay (RIA) procedures generally require high affinity antibodies whereas antibody immunoadsorbents for specific affinity chromatography require lower affinity antibodies. Thus it may be important at an early stage to use an assay that identifies MCA of particular affinity. MCA to be used finally in complement-dependent assays should be identified initially by their complement fixing ability which is restricted to certain Ig classes and subclasses. MCA are best detected in direct binding assays because of their class-restricted biological properties and their inability to precipitate antigens that do not have repeating epitopes. MCA binding to antigen is detected by a secondary agent such as an anti-globulin antiserum, Clq, or staphylococcal protein A (SpA), any one of which can be labelled with radioiodine, an enzyme, biotin or a fluorochrome. Direct binding assays can detect MCA of particular isotypes by employing an intermediate class-specific developing antiserum.

Solid phase assays are widely used and are readily adapted for soluble or particulate antigens, including whole cells, which can be linked to plastic tubes or multiwell plates. Enzyme-linked immunoadsorbent assay (ELISA) systems can handle large numbers of samples and are in many ways to be preferred to

RIA systems because they do not employ radioisotopes, are cheaper, can be done without expensive capital equipment. It is not possible to summarize here the great variety in the assays that can be used to detect MCA. Assays suitable for particular antigen–antibody systems are described in references given in the section on applications of the MCA technology and in the general collections of articles edited by Melchers *et al.* (1978), Kennett *et al.* (1980) and Hämmerling *et al.* (1981). In the early screening of growing hybridomas, accuracy of the assay may be sacrificed for speed, but it is necessary, if growing the wrong hybridomas is to be avoided, to identify at some stage those MCA that have the required specificity, assuming that the fusion was done with the aim of finding particular antibodies. Thus, precise specificity controls may need to be incorporated. For example, to identify MCA against polymorphic cell surface antigens, such as HLA, it is necessary to screen culture supernatants against a panel of HLA-typed cells. Further, many proteins, such as enzymes, not normally thought of as extensively polymorphic, have a micropolymorphism which is revealed by MCA directed against their variant parts which clearly influence the ease with which specific MCA can be detected (see, for example, Slaughter *et al.*, 1980). Rodent MCA discriminate very well between polymorphic variants of proteins from other species and while this may present problems in identifying MCA in the first place, once they have been isolated they may be used as probes in the fine genetic analysis of polymorphism. For use in assays, several such MCA may need to be combined to produce a standardized reagent that will bind specifically to all the polymorphic variants of a particular protein. Particular problems are encountered if the immunizing antigen was impure in the first place and purified material is not available to use in the culture supernatant screening assay. In such cases antibodies against the specific antigen and against contaminants cannot be differentiated from each other. There may be no alternative, therefore, to cloning many cultures with undifferentiated antibody activity and determining subsequently the specificity of the MCA secreted by the cloned cells. For MCA against complex soluble material the identity of the antigenic components with which they react can be found by using blotting techniques (Burnette, 1981). Antigen is electrophoresed in a thin-layer polyacrylamide gel and then electrophoretically transferred onto a nitrocellulose sheet and different channels exposed to individual MCA. Those antigen components that bind MCA can be identified by staining with an anti-Ig reagent. This can be either radioiodinated or linked to an enzyme such as horesradish peroxidase, in which cases the active materials are localized by autoradiography or immobilized enzyme assay respectively.

If MCA specific for a compound that has biological activity, but is otherwise unrecognizable, are required it will be necessary to screen the cell cultures for antibody that will neutralize or potentiate the biological activity of the compound. In this way Secher and Burke (1980) isolated MCA specific for interferon by exploiting the ability of antibody to inhibit the virus neutralizing activity of the interferon. Such bioassays take several days and therefore complicate the isolation of the specific antibody-secreting clones. In this case, however, the effort was repaid: one MCA was employed as an affinity immunoadsorbent and in a single chromatographic step isolated interferon from crude cell culture fluid at a high purity, representing a purification of some 5000 × over the starting material. Many MCA of other specificities have been isolated using bioassays rather than conventional serological assays. In any event, this may be a prerequisite for antibodies against enzymes, hormones or hormone receptors where MCA which modify the biological activity of the molecule are required in preference to those that merely bind to it.

The MCA technology should only be applied with caution in attempts to make specific antibodies if an assay system is not available to detect unequivocally antibodies of the required specificity.

6.2.8 Growing Hybrid Cell Lines and Harvesting MCA

By the very process of their selection, hybridomas are usually well adapted to grow *in vitro*. If conventional TC media supplemented with sera are used the spent medium will contain typically 1–10 μg/ml specific MCA if the hybridoma cells have been allowed to grow to exhaustion. This yield can be increased up to five-fold by repeatedly removing cells from the culture, thereby maintaining logarithmic growth for longer.

When MCA derived in this way are used for experiment or assay, the presence of large amounts of heterologous serum protein in the spent TC media may be undesirable for both technical and economic reasons. Hybridomas can be adapted to grow in low concentrations of serum, sometimes less that 1%, by gradually reducing the serum concentration over several days or weeks of culture. The use of Iscove's medium, which, if supplemented correctly will support the growth of some hybridomas in serum-free conditions, might be considered for this (Iscove and Melchers, 1978, Chang *et al.*, 1980).

Most cloned hybrid lymphocytes will grow as tumours in appropriate animals – a property they share with the parent myeloma lines – hence the derivation of the word hybridoma. Cells are implanted subcutaneously or intraperitoneally. The dose of cells required for a successful 'take', often initially in excess of 10^7, will decrease if the cell line becomes adapted for *in vivo* growth. It must be checked frequently for antibody secretion whether it is maintained either by *in vivo* passage or by *in vitro* culture.

Hybridoma cells form a palpable tumour within 2 weeks of subcutaneous implantation and antibody accumulates in the blood. Hybridomas isolated early after fusion often grow better subcutaneously rather than intraperitoneally. However, most lines when implanted in the latter site induce accumulation of ascites fluid containing MCA which will also be present at a similar level in the serum. Ascites fluid accumulation and intraperitoneal tumour growth can be enhanced by priming mice with light mineral oil (two intraperitoneal injections of 0.5 ml each, 3 weeks apart). Cells are implanted either intraperitoneally or subcutaneously when oil-induced fluid starts to accumulate in the peritoneal cavity (Potter, 1972).

MCA can reach concentrations of 10 mg/ml in ascites fluid and repeated drainage of the peritoneal cavity can provide more than 10 ml fluid per mouse. The obvious advantage of growing a hybridoma *in vivo* is the amount of antibody produced, but this will be contaminated with normal mouse Ig which may not be acceptable in all situations. Hybridomas will only grow either in animals which cannot reject them or in which they are histocompatible. Thus intraspecies hybrid animals are used to grow hybridomas derived from immune lymphocytes of a strain different from the myeloma parent. Interspecies hybridomas (e.g. rat × mouse) can be grown in nude mice or in either parental species immunosuppressed by anti-lymphocyte serum (McKearn, 1980). The rat offers distinct advantages, for 100 ml of ascites fluid can be obtained from a single animal. Athymic nude mice are suitable hosts for probably all hybridomas. Total histocompatibility may not be essential, but matching for *H-2* in normal mice should be achieved. Thus, (NZB × NZW)F_1 lymphocytes × BALB/c myeloma cell hybridomas will grow in (BALB/c × NZW)F_1 mice because NZB and BALB/c mice both carry the $H\text{-}2^d$ haplotype.

In many applications MCA need not be recovered from culture media or ascites fluids but if purification is required then several procedures are available for this. Standard methods for Ig purification are described by, for example, Weir (1978) and Lefkovits and Pernis (1979). Methods involving fractional salt precipitation and ion exchange chromatography were developed for isolating serum Ig which is of course highly heterogeneous and are not uniformly applicable to isolating every MCA. It is suggested that such methods are optimized for each MCA that is to be isolated for many do not behave in a conventional way. Preparative electrophoresis and isoelectric focusing are useful techniques and, if of the appropriate class, MCA can be recovered by affinity chromatography on staphylococcal protein A (Ey *et al.*, 1978; Goding, 1978) which requires relatively mild conditions for elution. Because of the usefulness of this technique for purifying MCA it may be worthwhile to use a screening assay (Section 6.2.7) based on the ability of the specific MCA to bind protein A (Jonsson and Kronvall, 1974). It must be emphasized that antibodies of IgM class and some IgG subclasses do not bind to protein A. Moreover, not all MCA of one subclass bind equally well. Whether this indicates the existence of Ig sub-subclasses is not known.

Affinity chromatography on anti-Ig coated gels can be used to purify MCA (and contaminating non-antibody Ig) and the elution conditions can be made less damaging by using anti-Ig antibodies of only low affinity. Alternatively, anti-Ig antibodies of high affinity can be blocked irreversibly by first exposing the immunoadsorbent to normal serum Ig.

Because of the high concentration of MCA relative to other immunoglobulins in both culture media and ascites fluids, techniques normally used for isolating Ig can be used here to isolate the specific antibody in good yield and purity. It will, however, contain copurified non-antibody Ig. The only way this can be avoided is by using affinity chromatography on a specific antigen immunoadsorbent. Elution procedures may be particularly abusive to both antibody and antigen and this approach cannot be applied universally. General discussion on purifying MCA can be found in the collected works and reviews quoted previously but see especially Lefkovits and Pernis (1979), Goding (1980), Kennett *et al.* (1980) and Galfré and Milstein (1981).

The conditions for storing MCA need to be determined experimentally. MCA can in many regards be treated like serum antibodies but some are almost totally destroyed or form aggregates after one cycle of freezing and thawing. It is recommended that MCA are stored sterile in concentrated form, if possible in the presence of carrier protein, either at 4 °C with a suitable preservative or at −20 °C or −70 °C. Repeated freezing and thawing should be avoided and once frozen, MCA should be maintained at the same storage temperature.

6.2.9 Labelling MCA

MCA can be labelled directly with fluorescein or an enzyme (e.g. horseradish peroxidase or alkaline phosphatase) by conventional methods. External radiolabelling with ^{125}I can also be achieved with the Chloramine-T method. These are widely discussed in the scientific literature but see especially Langone and Van Vunakis (1981).

MCA have one considerable advantage over conventional serum antibodies in that they can be easily labelled internally with one or a number of ^{3}H- or ^{14}C-labelled amino acids by incubating the hybridoma cells for up to 24 hours in an appropriate TC medium containing the amino acid(s) (Galfré and Milstein, 1981). Such antibodies are particularly useful for immunochemical and immunocytochemical procedures.

6.3 Homogeneity and Specificity of MCA

Although products of clones of cells, individual MCA are not structurally homogeneous. The native molecules commonly show multiple protein bands on polyacrylamide gel electrophoresis or on isoelectric focusing. Separated H chains can also show electrophoretically different forms. The chemical basis of these variations probably lies mainly in post-translational modification and in variations in the extent of glycosylation of the H chains of the molecules. To what extent these affect the functional specificity and biological properties of the MCA is not known. Mutation in the Ig genes and loss of chromosomes expressing Ig genes will occur within one clone and will also contribute to the inhomogeneity of MCA (see Section 6.2.3). MCA secreted by cells expressing Ig genes of the myeloma parent cell will be variously composed of antibody and myeloma polypeptide chains. For these reasons, and as stressed before, hybridomas and their secreted MCA must be checked regularly for somatic variations and the cells themselves must be preserved as a seed stock.

Some confusion may arise over the specificity of MCA. It should not, however, be a matter of surprise that some MCA cross-react with antigens other than those they were induced and selected against. They are, after all, only single representatives of the oligoclonal antibodies in serum and cross-reactions of antisera are well known. Antibody specificity can be defined at a molecular level by the molecules with which an antibody reacts. Because of the variety of epitopes on (e.g.) protein antigens it is possible to produce many MCA that bind to and are specific for the particular protein molecule in question. For example Reth *et al.* (1979) were able to produce many different anti-idiotopic MCA all reactive with the idiotopes of one MCA (used itself as the antigen in this situation). At the submolecular level an epitope or antigenic determinant is relatively small and stereochemical considerations determine whether antibody will bind effectively to it. Clearly then, an MCA will bind to any molecule with the correct stereochemistry. Accordingly, the cross-reactivity of an MCA reagent for an irrelevant antigen can never be removed by absorption to produce a specific reagent. The cross-reactivity is a property of all the MCA molecules. Wherever necessary MCA should be selected with this in mind: if it is undesirable to use MCA that cross-react with related or unrelated antigens then the screening process (Section 6.2.7) should account for this.

The affinity of an antibody is closely related to, and influences, its specificity. The functional affinity (Steward, 1981) of an MCA may be lower than that of the corresponding specific antiserum. This is likely to be due in part to the relatively low maximum antibody:antigen ratio that can be achieved in immune complexes formed between one MCA and an antigen that does not have repeating epitopes. Functional affinity can be increased by mixing together different MCA and this procedure may also lead to the formation of immune complexes that are large enough to precipitate from solution. This has been exploited by Jefferis *et al.* (1980) in the quantitation of serum levels of IgG of different light chain isotypes.

6.4 Uses and Applications of MCA

It is obvious that MCA could be used in many different situations especially those which exploit their purity, homogeneity, specificity and availability in almost unlimited amounts. Antisera prepared by conventional means are being replaced in many applications by MCA and no doubt this will continue. Antisera will, however, always be important not only because they are simple and relatively cheap to produce but also because they often have high functional affinity and an operationally precise specificity which arise from their oligoclonality.

The areas in which MCA are currently being used and those where they might find applications have been the subject of numerous reviews. Some areas have been touched upon previously in this chapter and discussion of the others is outside the scope of it.

General discussion of the uses and applications of MCA can be found in reviews by Milstein (1980), Hadden (1980), Yelton and Scharff (1980, 1981), Staines and Lew (1980), and Galfré and Milstein (1981). The collected papers edited by Melchers *et al.* (1978), Kennett *et al.* (1980), Hämmerling *et al.* (1981), and McMichael and Fabre (1982) cover most areas and should be consulted in conjunction with others listed below.

Particular areas of importance (with appropriate reviews and papers) are as follows:

(1) Identification and classification of major histocompatibility complex products for tissue typing in transplantation and population genetics studies (Möller, 1979; Zola, 1980; Eisenbarth, 1981).
(2) Identification of polymorphic molecules on cell surfaces in general and markers of cellular differentiation on lymphocytes (Kennett *et al.*, 1980; Milstein and Lennox, 1980; Raschke, 1980; Zola, 1980; Eisenbarth, 1981; Yelton and Scharff, 1981; and *Transplantation Proceedings* **12(3)**, 1980) and neural cells (Barnstable, 1982) in particular.
(3) Studies of tumour cell antigens, in particular the uses of MCA therapeutically and in body scanning (Boman and Fathman, 1981; Mach *et al.*, 1981).
(4) Serogenetic classification of infectious micro-organisms and protozoan and metazoan parasites, antigen purification for vaccine production and passive immunotherapy of infectious diseases (Rowe, 1980; Mitchell, 1981; Scharff *et al.*, 1981).
(5) Application of MCA in clinical laboratory test systems such as RIA and ELISA and the development of immunochemically based assays to replace conventional (e.g.) tissue section immunofluorescence assays. The opportunities here for standardization of assays are impressive. (Diamond *et al.*, 1981; Sevier *et al.*, 1981; McMichael and Fabre, 1982). MCA will also prove useful in forensic science laboratories (Fletcher and Davie, 1980).

In these areas it is clear that MCA will become well established as diagnostic agents. Much current research impinges upon this and upon the more prospective applications of MCA in the specific immunological manipulation of selected immune responses. Thus, by exploiting the specificity of MCA in recognizing particular antibody idiotypes (idiotopes) it may become possible either to suppress chosen immune responses as might be required in autoimmunity or tissue transplantation or to elevate them as required in tumour therapy and prophylactic immunization in general. Our ability to produce biological molecules of a chosen and well defined structure opens up the prospect of producing a new generation of molecules (which might have other active compounds linked to them) with binding specificity for any target organ or molecule in the body. Antibodies specific for hormones, hormone receptors or enzymes could become therapeutic agents for regulating other diseases which have no relationship to the immune system in the first place.

Acknowledgments

The author wishes to thank Dr. Felicity Grainger for help in preparing the manuscript, and, in particular, for providing details of information retrieval strategies.

Appendix

Strategies of Information Retrieval for Monoclonal Antibodies Using Medlars

'Antibodies, monoclonal' has been a Medical Subject Heading (MeSH) only since January 1982.

Publications indexed since that date can be searched in the hard copy Index Medicus and on Medline using this heading as an indexing term (via BLAISE-LINK) or as a descriptor (via DIALOG or DATASTAR).

Publications on monoclonal antibodies indexed earlier than 1982 can be found in the hard copy Index Medicus under the general heading 'Antibodies' and also under the more specific headings 'Antibodies, neoplasm' and 'Antibodies, heterotypic'. For searching Medline, a combination of MeSH as indexing term (via BLAISE-LINK) or descriptor (via DIALOG or DATASTAR) and text words can be used for pre-1982 material.

For example:

(1) antibodies or antibodies, neoplasm or antibodies, heterotypic (as indexing terms or descriptors);
(2) monoclonal or monospecific (as free words);
(3) 1 and 2.

These combinations can be used with other specific indexing terms to find information about monoclonal antibodies against defined or selected groups of antigens.

Glossary of Terms

B-lymphocyte A cell which synthesizes antibody molecules. Characteristically matures from precursors in the bone marrow without passing through the thymus. Also called B-cell.

Class Immunoglobulin type as defined by the structure of the heavy (H) chain. Subclasses are smaller taxonomic divisions of classes.

Clone Population or group of individuals (whole organisms or cells) all genetically identical, derived by somatic reproduction.

Dikaryon A cell with the nuclear material of two other somatic cells combined.

Epitope The part of an antigen molecule to which an antibody will bind. (Syn. = antigenic determinant).

Feeder cells Cells added to tissue cultures to facilitate the growth of other cells, particularly when these are at low concentration. Mode of action not completely understood but may provide nutritional metabolites.

H-2 The histocompatibility-2 gene system of the mouse which has major role in regulating immune responses. A system of closely linked genes, combinations of which (haplotypes) are identified, thus $H\text{-}2^a$, $H\text{-}2^b$, etc. Highly polymorphic products.

Hapten A molecule or chemical grouping that has the ability to combine with specific antibody against it, cannot induce an immune response on its own usually by virtue of its small size. Haptens usually induce immune responses when coupled to another molecule, thereby increasing their size and chemical complexity. Can be thought of as an isolated epitope.

Histocompatibility The situation where tissues can coexist without damaging each other. Applied to transplantation thus: histocompatible tissues are not rejected by the host onto which they are transplanted; histocompatibility genes code for histocompatibility antigens which are the components of transplanted tissues recognized and reacted to by the host's immune system.

HLA The human homologue of the mouse *H-2* system.

Hybridize To form into one the genetic material from two individual cells (or organisms). May be the result of combination of germ cells (sexual hybridization) or body cells (somatic hybridization).

Hybridoma A malignant tumour derived from the somatic hybridization of two (or more) other cells. Generally also used to describe cells (with a similar type of parentage) that will grow continuously *in vitro* (i.e. their malignancy not necessarily established).

Idiotype That part of an antibody molecule which makes it unique from other antibodies. Includes the part of the molecule which combines with antigen. Made up of individually smaller parts called idiotopes which can be defined in turn by anti-idiotopic antibodies against them.

Immunoadsorbent A particulate or immobilized antibody or antigen to which, respectively, specific antigen or antibody can be bound. Used extensively to remove unwanted antibodies from antisera and to purify specific antibodies by affinity chromatography.

Immunogen A substance that induces an immune response.

Immunogenicity The property of an antigen that describes its ability to induce an immune response.

Immunoglobulin (Ig) Protein produced by plasma cells and B-lymphocytes. Secreted and found in body fluids. Characteristic structure. Antibodies are immunoglobulins. In practice not all immunoglobulins can be identified as antibodies.

Isotype Structural type of the constituent immunoglobulin heavy and light polypeptide chains.

Monoclonal antibody (MCA) The antibody produced by one cell or a clone derived from it.

Myeloma Malignant tumour of bone marrow origin and of B-lymphocyte type. Usually secretes a myeloma protein which is an immunoglobulin.

Nude mice Animals carrying the *nu* (nude) gene. Homozygous *nu/nu* mice have very little fur and lack a thymus. In consequence of the latter they do not reject tissue transplanted from other mice or other species.

Oligoclonal Relating to several different clones.

Plasmacytoma Malignant tumour derived from plasma cell lineage. Usually secretes an immunoglobulin. Similar to myeloma but may be derived from ontogenetically more mature cell.

Rosetting technique Procedure whereby lymphocytes reactive with a specific antigen can be identified by their ability to bind to their surface an indicator (such as a foreign erythrocyte) particle to which the antigen has been chemically linked.

Seed stock Literally grain stored for next year's sowing. In this case, cells stored against the death or loss of function of existing continuous cell lines.

Somatic cell hybrid See: hybridize.

Specificity The property of an antibody (or antiserum) defined by its tendency to combine with a defined substance.

Syngeneic Of identical genetic constitution.

T-lymphocyte A cell derived from the bone marrow which matures only if it passes through the thymus. Does not secrete antibody. T-lymphocytes are of different types, classified according to their function: some regulate the activity of B-lymphocytes whereas others are involved in destruction of virus-infected cells or the generation of immunity against intracellular parasitic bacteria like *Mycobacterium*. Also called T-cell.

Tolerogen A substance that can be administered to an animal in such a way that the animal does not make an immune response against it upon subsequent re-exposure in conditions which evoke an immune response in an untreated animal.

Abbreviations

Clq	Component lq of the serum complement system
DMSO	Dimethylsulphoxide
EBV	Epstein–Barr virus
ELISA	Enzyme-linked immunosorbent assay
FCA	Freund's Complete Adjuvant
FCS	Fetal calf serum
FIA	Freund's Incomplete Adjuvant
H	Heavy chain (of immunoglobulin molecule)
HAT	Hypoxanthine, aminopterin and thymidine
HGPRT	Hypoxanthine guanine phosphoribosyl transferase
$HGPRT^+$	A cell expressing the HGPRT gene
$HGPRT^-$	A cell not expressing the HGPRT gene
HT	Hypoxanthine and thymidine
Ig	Immunoglobulin (IgG, IgM: immunoglobulin classes G and M respectively)
L	Light chain (of immunoglobulin molecule)
MCA	Monoclonal antibody(ies)
PBS	Phosphate buffered saline
PEG	Polyethylene glycol
RIA	Radioimmunoassay
SpA	Staphylococcal protein A
TC	Tissue culture

References

Andersson, J., and Melchers, F. (1978) The antibody repertoire of hybrid cell lines obtained by fusion of X63-AG8 myeloma cells with mitogen-activated B-cell blasts. *Curr. Top. Microbiol. Immunol.* **81**, 130–139

Anon (1976) Immunological adjuvants. *WHO Tech. Rep. Ser.* **595**

Barnstable (1982) Monoclonal antibodies – tools to dissect the nervous system. *Immunol. Today* **3**, 157–168

Boman, B. M., and Fathman, C. G. (1981) Monoclonal antibodies: the next attempt at tumour immunotherapy. *Mayo Clin. Proc.* **56**, 641–644

Boumsell, L., and Bernard, A. (1980) High efficiency of Biozzi's high responder mouse strain in the generation of antibody secreting hybridomas. *J. Immunol. Methods* **38**, 225–229

Burnette, W. N. (1981) Western blotting: electrophoretic transfer of proteins from sodium dodecyl sulfate-polyacrylamide gels to unmodified nitrocellulose and radiographic detection with antibody and radioiodinated protein A. *Anal. Biochem.* **112**, 195–203

Buttin, G., LeGuern, G., Phalente, L., Lin, E. C. C., Medrano, L., and Cazenave, P. A. (1978) Production of hybrid lines secreting monoclonal anti-idiotypic antibodies by cell fusion on membrane filters. *Curr. Top. Microbiol. Immunol.* **81**, 27–36

Chang, T. H., Steplewski, Z., and Koprowski, H. L. (1980) Production of monoclonal antibodies in serum free medium. *J. Immunol. Methods* **39**, 369–375

Coffino, P., and Scharff, M. D. (1971) Rate of somatic mutation in immunoglobulin production by mouse myeloma cells. *Proc. Nat. Acad. Sci. USA* **68**, 219–223

Cook, W. D., and Scharff, M. D. (1977) Antigen-binding mutants of mouse myeloma cells. *Proc. Nat. Acad. Sci. USA* **74**, 5687–5691

Cotton, R. G. H., and Milstein, C. (1973) Fusion of two immunoglobulin-producing myeloma cells. *Nature (London)* **244**, 42–43

Cotton, R. G. H., Secher, D., and Milstein, C. (1973) Somatic mutation and the origin of antibody diversity. Clonal variability of the immunoglobulin produced by MOPC 21 cells in culture. *Eur. J. Immunol.* **3**, 135–140

Croce, C. M., Linnenbach, A., Dolby, T. W., and Koprowski, H. (1981) In: *Monoclonal Antibodies and T-cell Hybridomas. Perspectives and Technical Advances* (eds G. J. Hämmerling, U. Hämmerling and J. F. Kearney), pp. 432–444. Elsevier/North-Holland Biomedical Press, Amsterdam

Croce, C. M., Linnenbach, A., Hall, W., Steplewski, Z., and Koprowski, H. (1980) Production of human hybridomas secreting antibodies to measles virus. *Nature (London)* **288**, 488–489

Davidson, R. L., and Ephrussi, B. (1970) Factors influencing the 'effective mating rate' of mammalian cells. *Exp. Cell Res.* **61**, 222–226

Davidson, R. L., and Gerald, P. S. (1976) Improved techniques for the induction of mammalian cell hybridization by polyethylene glycol. *Somatic Cell Genetics* **2**, 165–176

Diamond, B. A., Yelton, D. E., and Scharff, M. D. (1981) Monoclonal antibodies. A new technology for producing serologic reagents. *New Engl. J. Med.* **304**, 1344–1349

Eisenbarth, G. S. (1981) Application of monoclonal antibody techniques to biochemical research. *Anal. Biochem.* **111**, 1–16

Ephrussi, B. (1972) *Hybridization of Somatic Cells.* Princeton University Press, Princeton, NJ, USA

Ey, P. L., Prowse, S. J., and Jenkin, C. R. (1978) Isolation of pure IgG_1, IgG_{2a} and IgG_{2b} immunoglobulins from mouse serum using protein A-sepharose. *Immunochem.* **15**, 429–436

Fazekas de St. Groth, S., and Scheidegger, D. (1980) Production of monoclonal antibodies: strategy and tactics. *J. Immunol. Methods* **35**, 1–21

Fletcher, S. M., and Davie, M. J. (1980) Monoclonal antibody: a major new development in immunology. *J. Forens. Sci. Soc.* **20**, 163–167

Galfré, G., Butcher, G. W., Howard, J. C., Wilde, C. D., and Milstein, C. (1980) Clonal competition and stability of hybrid myelomas of mouse and rat origin. *Transplant. Proc.* **12**, 371–375

Galfré, G., Howe, S. C., Milstein, G., Butcher, G. W., and Howard, J. C. (1977) Antibodies to major histocompatibility antigens produced by hybrid cell lines. *Nature (London)* **266**, 550–552

Galfré, G., and Milstein, C. (1981) Preparation of monocloncal antibodies: strategies and procedures. *Methods Enzymol.* **73B**, 1–46
Galfré, G., Milstein, C., and Wright, B. (1979) Rat × rat hybrid myelomas and a monoclonal anti-F_d portion of mouse IgG. *Nature (London)* **277**, 131–133
Goding, J. W. (1978) Use of staphylococcal protein A as an immunological reagent. *J. Immunol. Methods* **20**, 241–253
Goding, J. W. (1980) Antibody production by hybridomas. *J. Immunol. Methods* **39**, 285–308
Hadden, J. W. (1980) Hybridoma antibodies. *Clin. Bull.* **10**, 26–29
Hämmerling, G., Hämmerling, U., and Kearney, J. F. (eds) (1981) *Monoclonal Antibodies and T-cell Hybridomas. Perspectives and Technical Advances.* Elsevier/North-Holland Biomedical Press, Amsterdam
Horibata, K., and Harris, A. W. (1970) Mouse myelomas and lymphomas in culture. *Exp. Cell Res.* **60**, 61–77
Iscove, N. N., and Melchers, F. (1978) Complete replacement of serum by albumin, transferrin, and soybean lipid in cultures of lipopolysaccharide-reactive B-lymphocytes. *J. Exp. Med.* **147**, 923–933
Jefferis, R., Deverill, I., Ling, N. R., and Reeves, W. G. (1980) Quantitation of human total IgG, kappa IgG and lambda IgG in serum using monoclonal antibodies. *J. Immunol. Methods* **39**, 355–362
Jonsson, S., and Kronvall, G. (1974) The use of protein A-containing *Staphyloccus aureus* as a solid phase anti-IgG reagent in radioimmunoassays as exemplified in the quantitation of α-fetoprotein in normal human adult serum. *Eur. J. Immunol.* **4**, 29–33
Kearney, J. F., Radbruch, A., Liesegang, B., and Rajewsky, K. (1979) A new mouse myeloma cell line that has lost immunoglobulin expression but permits the construction of antibody-secreting hybrid cell lines. *J. Immunol.* **123**, 1548–1550
Kennett, R. H., and Gilbert, F. (1979) Hybrid myelomas producing antibodies against a human neuroblastoma antigen present on fetal brain. *Science* **203**, 1120–1121
Kennett, R. H., McKearn, T. J., and Bechtol, K. B. (eds) (1980) *Monoclonal Antibodies: Hybridomas: A new Dimension in Biological Analyses.* Plenum Press, New York
Kenny, P. A., McCaskill, A. C., and Boyle, W. (1981) Enrichment and expansion of specific antibody-forming cells by adoptive transfer and clustering, and their use in hybridoma production. *Aust. J. Exp. Biol. Med. Sci.* **59**, 427–437
Köhler, G., and Milstein, C. (1975) Continuous cultures of fused cells secreting antibodies of predefined specificity. *Nature (London)* **256**, 495–497
Köhler, G., and Milstein, C. (1976) Derivation of specific antibody-producing tissue culture and tumour lines by cell fusion. *Eur. J. Immunol.* **6**, 511–519
Kranz, D. M., Billing, P. A., Herron, J. N., and Voss, E. W. (1980) Modified hybridoma methodology: antigen-directed chemically mediated cell fusion. *Immunol. Comm.* **9**, 639–651
Langone, J. J., and Van Vunakis, H. (eds) (1981) *Immunochemical Techniques. Methods in Enzymology* **73(B)**. Academic Press, New York.
Lefkovits, I., and Pernis, B. (eds) (1979) *Immunological Methods*, Vol. I. Academic Press, New York
Lefkovits, I., and Waldmann, H. (1979) *Limiting Dilution Analysis of cells in the Immune System*. Cambridge University Press, Cambridge
Littlefield, J. W. (1964) Selection of hybrids from matings of fibroblasts *in vitro* and their presumed recombinants. *Science* **145**, 709–710
Mach, J.-P., Buchegger, F., Forni, M., Ritschard, J., Berche, C., Lumbroso, J.-D., Schreyer, M., Girardet, C., Accolla, R. S., and Carrel, S. (1981) Use of radiolabelled

monocloncal anti-CEA antibodies for the detection of human carcinomas by external photoscanning and tomoscintigraphy. *Immunol. Today* **2**, 239–249

Margulies, D. H., Kuehl, W. M., and Scharff, M. D. (1976) Somatic cell hybridization of mouse myeloma cells. *Cell* **8**, 405–415

McConnell, I., Munro, A., and Waldmann, H. (1981) *The Immune System. A Course on the Molecular and Cellular Basis of Immunity*, 2nd edn. Blackwell, Oxford

McKearn, T. J. (1980) In: *Monoclonal Antibodies: Hybridomas: A New Dimension in Biological Analyses* (eds R. H. Kennett, T. J. McKearn and K. B. Bechtol), pp. 403–404. Plenum, New York

McMichael, A., and Fabre, J. (eds) (1982) *Monoclonal Antibodies in Clinical Medicine*. Academic Press, London

Melchers, F., Potter, M., and Warner, N. L. (eds) (1978) *Lymphocyte Hybridomas. Curr. Top. Microbiol. Immunol.* **81**

Milstein, C. (1980) Monoclonal Antibodies. *Sci. American* **243**, 56–64

Milstein, C., and Lennox, E. (1980) The use of monoclonal antibody techniques in the study of developing cell surfaces. *Curr. Top. Develop. Biol.* **14**, 1–32

Mitchell, G. F. (1981) Hybridoma antibodies in immunodiagnosis of parasite infection. *Immunol. Today* **2**, 140–142

Möller, G. (ed) (1979) *Immunol. Revs.* **47**

Oi, V. T., and Herzenberg, L. A. (1980) In: *Selected Methods in Cellular Immunology* (eds B. B. Mishell and S. M. Shiigi), pp. 351–372. W. H. Freeman, San Francisco

Olsson, L., and Kaplan, H. S. (1980) Human–human hybridomas producing monoclonal antibodies of predefined antigenic specificity. *Proc. Nat. Acad. Sci. USA* **77**, 5429–5431

Pages, J. M., and Bussard, A. E. (1978) Establishment and characterization of a permanent murine hybridoma secreting monoclonal autoantibodies. *Cell Immunol.* **41**, 188–194

Parks, D. R., Byran, V. M., Oi, V. T., and Herzenberg, L. A. (1979) Antigen-specific identification and cloning of hybridomas with a fluorescence-activated cell sorter. *Proc. Nat. Acad. Sci. USA* **76**, 1962–1966

Paul, J. (1975). *Cell and Tissue Culture*, 5th edn, E. and S. Livingstone, London

Pluznik, D. H., and Sachs, L. (1965) The cloning of normal 'mast' cells in tissue culture. *J. Cell Comp. Physiol.* **66**, 319–324

Pontecorvo, G. (1975) Production of mammalian somatic cell hybrids by means of polyethylene glycol treatment. *Somatic Cell Genetics* **1**, 397–400

Potter, M. (1972) Immunoglobulin-producing tumours and myeloma proteins of mice. *Physiol. Rev.* **52**, 631–719

Raschke, W. C. (1980) Plasmacytomas, lymphomas and hybridomas: their contribution to immunology and molecular biology. *Biochem. Biophys. Acta* **605**, 113–145

Reth, M., Imanishi-Kari, T., and Rajewsky, K. (1979) Analysis of the repertoire of anti-(4-hydroxy-3-ntrophenyl)acetyl (NP) antibodies in C57BL/6 mice by cell fusion. II. Characterization of idiotopes by monoclonal anti-idiotype antibodies. *Eur. J. Immunol.* **9**, 1004–1013

Ringertz, N. R., and Savage, R. E. (1976) *Cell Hybrids*. Academic Press, London

Rowe, D. S. (1980) The role of monoclonal antibody technology in immunoparasitology. *Immunol. Today* **1**, 30–33

Scharff, M. D., Roberts, S., and Thammana, P. (1981) Monoclonal antibodies. *J. Infect. Dis.* **143**, 346–351

Schneider, M. D., and Eisenbarth, G. S. (1979) Transfer plate radioassay using cell monolayers to detect anti-cell surface antibodies synthesized by lymphocyte hybridomas. *J. Immunol. Methods* **29**, 331–342

Schneiderman, S., Farber, J. L., and Baserga, R. (1979) A simple method for decreasing the toxicity of polyethylene glycol in mammalian cell hybridization. *Immunogenetics* **5**, 263–269

Schröder, J., Sutinen, M.-L., and Suomalainen, H. A. (1981) Chromosome segregation in lymphocyte hybrids. *Hereditas* **94**, 77–82

Schwaber, J. (1975) Immunoglobulin production by a human–mouse somatic cell hybrid. *Exp. Cell Res.* **93**, 343–354

Secher, D. S., and Burke, D. C. (1980) A monoclonal antibody for large-scale purification of human leukocyte interferon. *Nature (London)* **285**, 446–450

Sevier, E. D., David, G. S., Martinis, J., Desmond, W. J., Bartholomew, R. M. and Wang, R. (1981) Monoclonal antibodies in clinical immunology. *Clin. Chem.* **27**, 1797–1806

Sharon, J., Morrison, S. L., and Kabat, E. A. (1979) Detection of specific hybridoma clones by replica immunoadsorption of their secreted antibodies. *Proc. Nat. Acad. Sci. USA* **76**, 1420–1424

Shulman, M., Wilde, C. D., and Köhler, G. (1978) A better cell line for making hybridomas secreting specific antibodies. *Nature (London)* **276**, 269–270

Sinkovics, J. G. (1981) Early history of specific antibody-producing lymphocyte hybridomas. *Cancer Res.* **41**, 1246–1247

Slaughter, C. A., Coseo, M. C., Abrams, C., Cancro, M. P., and Harris, H. (1980) In: *Monoclonal Antibodies: Hybridomas: A New Dimension in Biological Analyses.* (eds R. H. Kennett, T. J. McKearn and K. B. Bechtol), pp. 103–120. Plenum, New York

Stähli, C., Staehelin, T., Miggiano, V., Schmidt, J., and Häring, P. (1980) High frequencies of antigen-specific hybridomas: dependence on immunization parameters and prediction by spleen cell analysis. *J. Immunol. Methods* **32**, 297–304

Staines, N. A., and Lew, A. M. (1980) Whither monoclonal antibodies? *Immunology* **40**, 287–293

Steinitz, M. (1981) In: *Monoclonal Antibodies and T-cell Hybridomas. Perspectives and Technical Advances* (eds G. J. Hämmerling, U. Hämmerling and J. F. Kearney), pp. 447–452. Elsevier/North-Holland, Amsterdam

Steinitz, M., Klein, G., Koskimies, S., and Mäkelä, O. (1977) EB virus-induced B lymphocyte lines producing specific antibody. *Nature (London)* **269**, 420–422

Steinitz, M., Koskimies, S., Klein, G. and Mäkelä, O. (1979) Establishment of specific antibody producing human lines by antigen preselection and Epstein–Barr virus (EBV) – transformation. *J. Clin. Lab. Immunol.* **2**, 1–7

Steward, M. W. (1981) The biological significance of antibody affinity. *Immunol. Today* **2**, 134–140

Szybalski, W., Szybalska, E. H., and Ragni, G. (1962) Genetic studies with human cell lines. *Natl Cancer Inst. Monogr.* **7**, 75–89

Trowbridge, I. (1978) Interspecies spleen-myeloma hybrid producing monoclonal antibodies against mouse lymphocyte surface glycoprotein, T200. *J. Exp. Med.* **148**, 313–323

Weir, D. M. (ed) (1978) *Handbook of Experimental Immunology,* 3rd edn, Blackwell, Oxford

Whitehouse, M. W. (1977) In: *Immunochemistry: An Advanced Textbook* (eds L. E. Glynn and M. W. Steward), pp. 571–601. John Wiley, Chichester

Yelton, D. E., and Scharff, M. D. (1980) Monoclonal antibodies. *Am. Sci.* **68**, 510–516

Yelton, D. E., and Scharff, M. D. (1981) Monocloncal antibodies: a powerful new tool in biology and medicine. *Ann. Rev. Biochem.* **50**, 657–680

Zola, H. (1980) Monoclonal antibodies against human cell membrane antigens: a review. *Pathology* **12**, 539–557

Biochemical Research Techniques
Edited by J. M. Wrigglesworth

7
Cell Culture

ALAN H. BITTLES

Department of Human Biology, Chelsea College, University of London, London SW3 6LX, UK

7.1 Introduction

The first successful experiments in maintaining tissue *in vitro* were reported by Roux and his co-workers in Berlin in 1885. By bathing portions of chick embryo in warm saline the tissue fragments could be kept alive in the laboratory for several days. However, it was the introduction of the plasma clot culture method by Burrows in 1910 and Carrel in 1912(b) that enabled tissue to be not only maintained but actually grown and so laid the foundations for the subsequent development of the technique. While organ culture, the method of culturing portions of intact tissue, is of considerable interest in many areas of anatomical and physiological research, the main application of tissue culture techniques in the biomedical field is dissociated cell culture and this will form the focal point of the present chapter.

Besides applications in areas such as virology and in providing valuable information on cellular organization by scanning and transmission electron microscopy (Figs 1 and 2), one of the greatest attractions of dissociated cell culture derives from the phenomenon of 'de-differentiation'. Cells such as fibroblasts that *in vivo* have limited function undergo de-repression *in vitro*, resulting in the expression of large portions of their genome. Thus fibroblasts cultured either from the skin or amniotic fluid can be used in the diagnosis and investigation of many inborn errors of metabolism.

In theory it is possible to grow nucleated cells from virtually any source in the body but in practice the success rate varies widely according to the nature of the

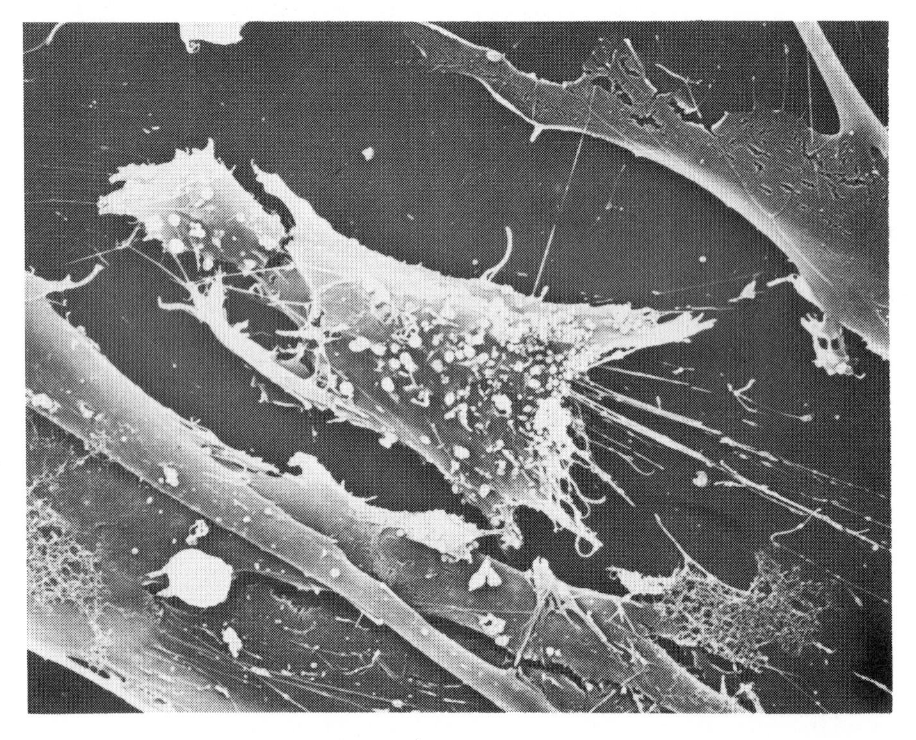

Fig. 1 Human diploid fibroblast in mitosis: population doubling level 23. Cells in mitosis are rounder and less flattened in appearance than their non-mitotic counterparts (Fig. 4). Blebs and microvilli are concentrated on the surface with filopodia stretching across the surrounding growth area. Scanning electron microscopy, magnification × 870. (Reproduced by kind permission of Dr. Yula Sambuy)

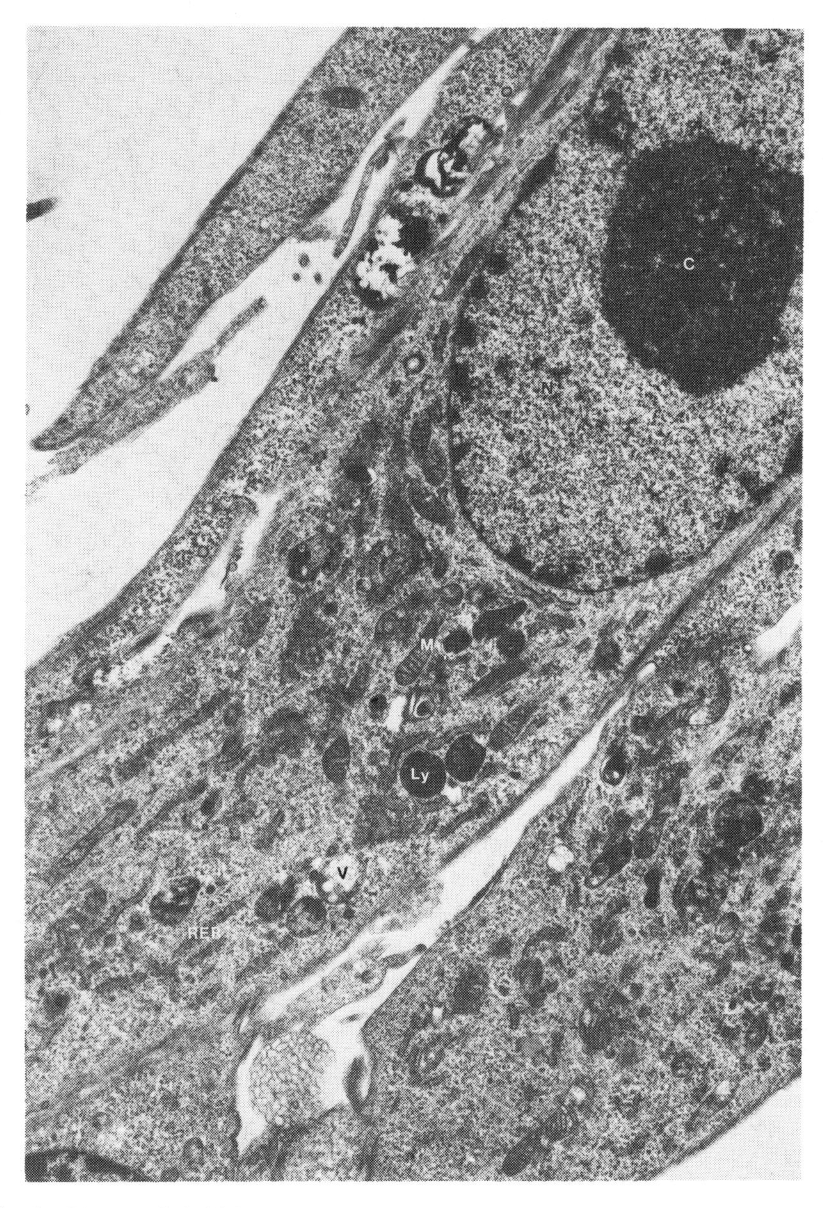

Fig. 2 Human diploid fibroblast: population doubling level 45. Transmission electron microscopy, magnification × 11,200. Key: N, nucleus; C, centriole; M, mitochondrion; Ly, lysosome; V, vacuole; RER, rough endoplasmic reticulum. (Reproduced by kind permission of Dr. Yula Sambuy)

tissue employed. The highest rates of success generally will be achieved with tissue which *in vivo* is continuously mitotic throughout the life span of the source species (Goldstein, 1971). Within this category cells derived from embryonic or very young donors are both the most rapidly established in culture and display the longest *in vitro* period of growth, the exact life-span being a characteristic of the species and, at least in human cell lines, inversely proportional to the age of the donor (Martin *et al.*, 1970). Cultures started from neoplastic tissue or which have been transformed *in vitro* to a heteroploid karyotype display markedly different growth characteristics to those of their diploid counterparts. These are summarized in Table 1.

Table 1 *In vitro* growth characteristics

Normal cell lines	Established (transformed) lines
Diploid karyotype	Abnormal chromosome number
Histologically normal	Abnormal histology
Finite growth period	Indefinite life span
Anchorage-dependent: require monolayer culture	Usually not anchorage-dependent: can be grown in suspension culture

It can be seen that the major limiting factors in using diploid cultures are:

(1) their finite growth period, often referred to as the 'Hayflick limit' and demonstrated in Fig. 3. Prior to the studies of Hayflick and Moorhead (1961) it had been believed that cells cultured *in vitro* had an indefinite life span (Carrel, 1912a) but this is now recognized to be a property only of transformed cell lines.
(2) their anchorage-dependence which requires the provision of a solid support medium.
(3) the phenomenon of contact or density-dependent inhibition displayed by diploid cells which means that they can only be grown as monolayer cultures, as in Fig. 4. By comparison, static cultures of transformed cells are observed to pile up several layers in depth as the result of a defect in this mechanism (Abercrombie, 1979).

7.2 Basic Equipment Requirements

Four basic pieces of equipment are required for cell culture. A *laminar flow cabinet* is needed for the basic manipulative procedures. The type of cabinet required will be dependent on the nature of the starting material but for non-malignant human, animal and avian tissue a downward displacement model equipped with internal filters and an ultraviolet source would suffice. On each occasion prior to usage, the working surfaces within the cabinet should be chemically cleansed using a suitable disinfectant such as Savlon and exposed to UV irradiation for a 10 minute period. It must be stressed that these measures are merely aids to aseptic handling techniques and not substitutes.

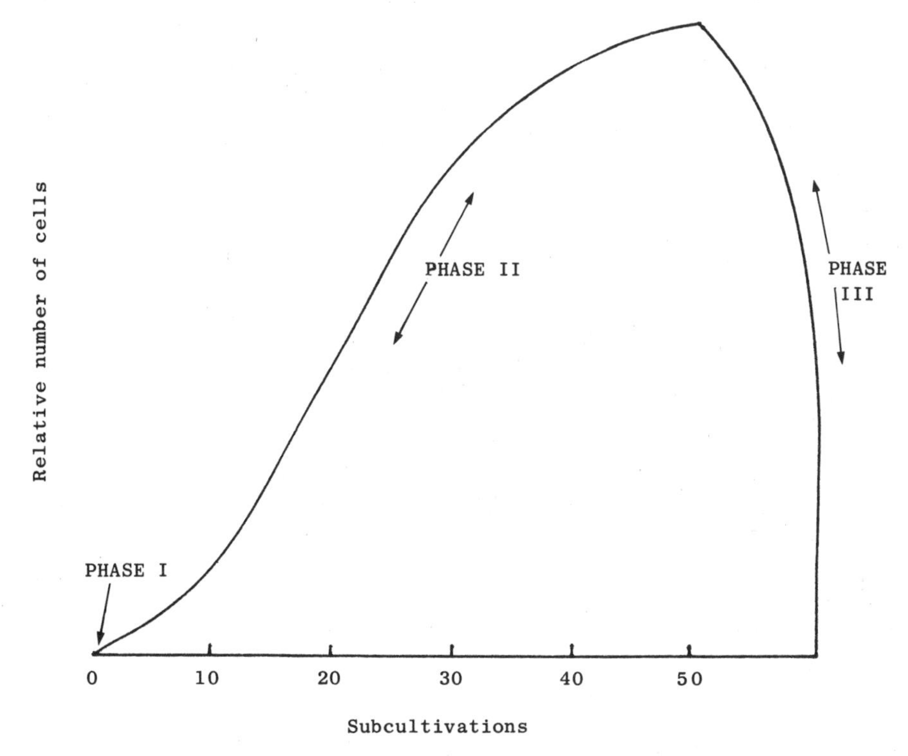

Fig. 3 Diagrammatic representation of the life cycle of a diploid cell line. Phase 1: the primary culture, ends with the first subcultivation. Phase 2: characterized by rapid growth and repeated subcultivations. Phase 3: the loss of growth potential of the culture, *in vitro* senescence. (From Hayflick and Moorhead (1961) by permission of Academic Press Inc.)

Most mammalian and avian cells have an optimal growth rate between 37.0°C and 38.5 °C and so an *incubator* capable of operating between these temperatures with a tolerance of ± 0.2 °C is necessary. A refrigerated incubator gives superior temperature control particularly where marked ambient temperature differences may exist. If culture systems based on petri dishes are required a carbon dioxide incubator is virtually obligatory and facilities for monitoring the supply of CO_2 to the cultures should be incorporated in the incubator design. As an alternative to a CO_2 incubator, sealed clear plastic boxes flushed with a

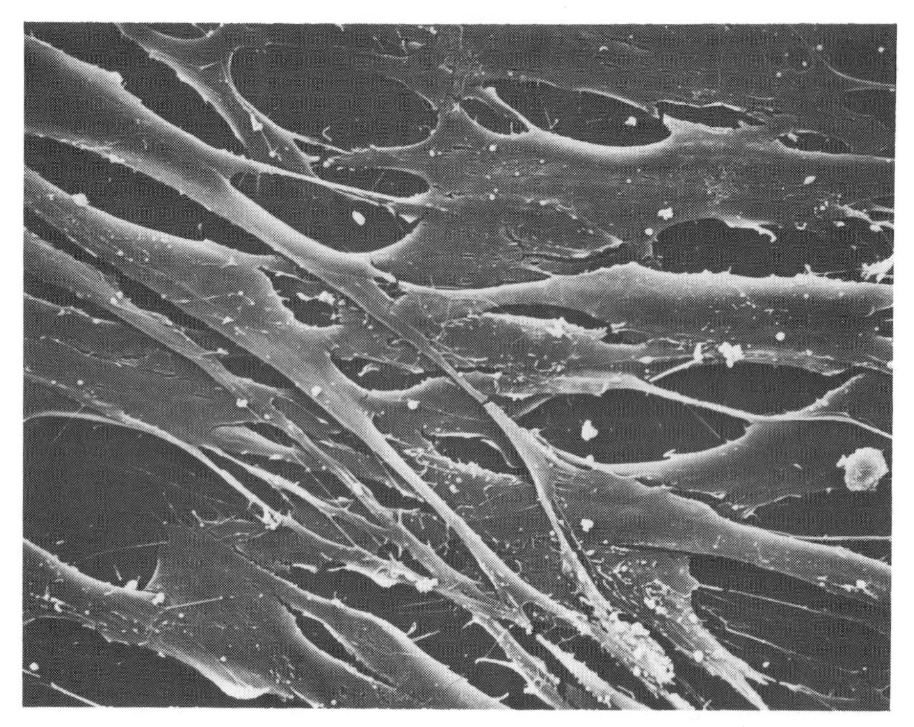

Fig. 4 Human diploid fibroblasts: population doubling level 23. Typical monolayer culture with a degree of overlapping between adjacent cells. There is considerable heterogeneity in size and shape, cell contact being maintained *via* the filopodia which may extend to several micrometers in length. Scanning electron microscopy, magnification × 870. (Reproduced by kind permission of Dr. Yula Sambuy)

5% CO_2/95% air mixture can be used on an *ad hoc* basis but are only feasible for small numbers of dishes undergoing short-term culture.

An *inverted microscope* capable of studying colonial morphology in adequate detail is necessary. For this purpose an instrument equipped with a 10 × eyepiece and two objectives of 4 × and 10 × is suitable. Ideally the microscope should be capable of examining as wide a variety of culture vessels as possible; these may range from petri dishes of 30 × 10 mm to roller bottles, 550 × 110 mm. Phase contrast facilities are a useful optional addition.

Unless sequential experiments are being carried out it is unnecessarily time-consuming and expensive to maintain cells under continuous cultivation and so a means of storage is required. This can be achieved either at −70 °C in a low temperature cabinet or with a *liquid nitrogen refrigerator*. The latter is preferable in many respects since it has the advantages of a lower storage temperature, is independent of the mains power supply and is more compact. Two types of liquid nitrogen refrigeration are available: compartment storage utilizing the vapour phase of the gas at −130 °C and the 'straw' method which employs canisters suspended in the liquid nitrogen with the correspondingly lower temperature of −196 °C. The major advantage of the compartment system is ease of storage and retrieval whereas the canister system has minimal gas loss by virtue of the narrower vessel neck. Maximum gas losses from the two types of refrigerators under static conditions are approximately 2.3 and 0.7 litres per day respectively. The provision of a roller base as an optional accessory is recommended for both systems as it allows the safe and easy movement of the nitrogen refrigerator in the laboratory.

7.3 Practical Aspects of Cell Culture

7.3.1 Tissue Preparation

The three main sources from which normal diploid cells are obtained for culture and subsequent biochemical assay are:

(1) skin or large organ material by biopsy;
(2) the peripheral circulation by venepuncture;
(3) amniotic fluid cells by amniocentesis.

In the latter two cases the preparation of the cells for culture is relatively straightforward as lymphocytes can be set up without pretreatment and stimulated to go into mitosis using phytohaemagglutinin while amnion cells may be isolated from the amniotic fluid by low speed centrifugation (not more than 60 *g*). Skin or organ biopsies, which should be taken into and stored in transport medium incorporating double strength antibiotics before processing, do require dissociation into suitably small explants of 0.5–1.0 mm^3 in order to initiate cell culture. This can be achieved by mechanical dissociation, for example with two scalpels worked in opposition to one another using clean cuts to minimize tearing of the tissue, and is particularly suitable for skin fibroblast culture when the initial sample has been taken by pinch biopsy (Edwards, 1960), with or without local anaesthesia. Alternatively a commercially available tissue dissociator can be employed which may be more applicable to the culture of explants derived from the biopsy of large organs, such as liver or kidney.

Enzyme dissociation with trypsin or collagenase is also appropriate for the preparation of large organ explants and can be most efficiently applied in combination with prior mechanical dissociation. The disadvantage to the use of enzymic dissociation is the necessity of halting the action of the enzyme after the requisite degree of tissue degradation has been completed and without employing conditions inimical to the tissue explants. The addition of fetal bovine serum to the tissue–enzyme digest is the simplest method of meeting this goal.

7.3.2 The Support Medium

As noted in Table 1, normal diploid cells are anchorage-dependent whereas transformed heteroploid cells are usually able to grow in suspension culture. In the 1940s and 1950s a bewildering array of natural materials was investigated as possible support media ranging from silk veil and spiders' webs to glass wool but not surprisingly their usefulness and availability was strictly limited. With the description by Sanford *et al.* and Evans *et al.* in 1951 of suitable glass culture vessels, monolayer culture could be commenced on a greatly increased scale and the subsequent commercial introduction of tissue culture grade disposable polystyrene petri dishes and flasks has further facilitated reproducibility in the initiation and routine maintenance of cultures. The most recently introduced support medium is the microcarrier system initially based on Sephadex beads and available in a variety of forms (see Section 7.3.3).

With all three of the major support media the basic factor governing the initial interaction between the cells with their negative surface charges (Borysenko and Woods, 1979) and the support medium would appear to be the density of the charges on the surface of the support rather than their polarity (Maroudas, 1975). The surfaces of the glass and polystyrene vessels are also negatively charged whereas the standard microcarriers carry a positive surface charge.

For virtually all cell lines tissue culture quality glass, whether borosilicate or soda, and polystyrene vessels are virtually identical in their abilities to act as growth supports although the polystyrene vessels usually give a proportionately greater cell yield because of their more uniform growth surface. When culturing in a glass vessel for the first time a poor cell yield may be obtained but the problem usually disappears on subsequent use. The exact cause is still somewhat obscure but may be associated with changes occurring in the charge distribution on the surface of the glass.

The general advantage of glass over polystyrene is that it is reusable but this presupposes the availability of specialized tissue culture wash-up and sterilizing facilities which are expensive both to install and run. The choice between the two is therefore based mainly on the scale of the culture facility and on convenience. For relatively small-scale culture units polystyrene vessels are best as they are cost-effective and the risks of using an inadequately cleaned or non-sterile culture vessel is minimal. The general advantage of the microcarriers lies in their mode of culture enabling diploid cells to be grown under simulated suspension conditions.

7.3.3 Types of Culture Available

Three main forms of culture are available for normal, diploid cells:

(1) static culture in microtitration plates, petri dishes or flasks;

(2) roller culture with bottles of varying dimensions;
(3) microcarrier culture.

Each has its advantages and disadvantages and the choice of culture system should be based mainly on the type of cell to be cultured and the nature of the experiment. The main advantages with microtitration and petri dish cultures are economy, convenience and the speed of establishment of the cultures. For those cells which are especially sensitive to changes in medium pH and gas partial pressure, these semi-open cultures can be optimally maintained in a CO_2 incubator but their culture on a long-term basis under these conditions would not be feasible because of the risk of contamination by micro-organisms. A general disadvantage associated with all static cultures lies in the suboptimal nutrient and waste product exchange between the cells and the growth medium, basically dependent on convection. Roller culture using glass or polystyrene bottles overcomes this problem and consequently many cell lines, both normal and transformed, show a faster growth rate and a higher cell yield indicating better growth conditions and a more efficient use of the medium. Roller culture also enables the scale of growth to be greatly increased which particularly in enzyme and cell product studies can be invaluable.

The microcarrier system for growing anchorage-dependent cells on DEAE-Sephadex A-50 beads kept in suspension by stirring was first devised by van Wezel in 1967. The beads had the necessary prerequisites for culture of a suitably charged culture surface in combination with a large surface area to volume ratio and their density meant that they could be kept in suspension without employing high stirring speeds. The system has since been developed commercially by several companies and for example, from one of them, Pharmacia, three microcarrier types are now available: Cytodex 1 has positively charged groups throughout the matrix and produces growth rates for most human and animal cell lines comparable to those obtained on glass or polystyrene; in Cytodex 2 the stable charged groups are restricted to the surface layer thus minimizing the binding of cell products to the microcarrier matrix, while Cytodex 3 has a surface layer of denatured collagen for the attachment of cells in place of the synthetic charged groups of the other two forms.

The optimum inoculation density for microcarrier culture varies between cell lines as it is dependent on their relative plating efficiencies. In effect the attachment efficiency of cells to the microcarriers is similar to that observed with petri dishes and so the microcarrier inoculum in cells/cm^2 can be calculated on this basis. A major technical decision with microcarrier cultures lies in choosing the method for maintaining the beads in suspension as this can have a significant effect on cell yield. Suspension has variously been achieved using a magnetic stirrer or a suspended stirring rod with rod stirring generally proving to be the more efficient. The speed of stirring is also important in determining the growth and final yield of cells. Excessively fast stirring speeds increase the shearing forces on the cells attached to microcarriers reducing the overall yield. Conversely cell yield may also be reduced by slow stirring with its resultant inefficient gas diffusion combined with aggregation and sedimentation of the microcarriers. To help prevent such aggregation all glass surfaces with which the microcarriers might come into contact should be siliconized before use, preferably with a mixture of dimethyldichlorosilane in an organic solvent. Included in this category are all pipettes and storage bottles as well as the culture vessels themselves.

Transformed cells usually do not exhibit anchorage-dependence and so can be grown directly by suspension culture. Cells grown in this manner often appear to require a period of adaptation and when first introduced into a suspension mode of growth a lag phase of several days may ensue before the commencement of rapid proliferation. Initially suspension cultures were established by the 'tumble-tube' method consisting of culture tubes arranged radially on a roller-drum. Later developments used rotary shaking to produce a swirling movement in the growth medium and currently the method of choice is magnetic stirring but as with microcarrier culture, a wide variety of stirring devices have also been successfully employed. The great advantage of suspension culture is the high yield that can be obtained with suitable cell lines and the option of establishing continuous cultures which further increase the efficiency of the process. Suspension cultures have a tendency to precipitate material, probably originating from the serum used as a growth supplement, on to the inner surface of the culture vessel. This can be prevented using methylcellulose or polyglycol. The major drawback to this form of culture is its restriction to heteroploid cell lines.

7.3.4 The Growth Medium

A wide range of growth media have been formulated and are commercially available for the *in vitro* culture of cells, their composition being based largely on classic depletion experiments carried out some 25 years ago (Eagle 1955, 1956; Eagle *et al.*, 1958, Eagle and Piez, 1960). They have in common a synthetic mixture of inorganic salts which maintain the pH and osmotic pressure of the culture and additionally act as an energy source, glucose as the main energy source, a range of amino acids and vitamins defined as essential for growth and a pH indicator solution, sodium phenol red. The media are usually available in three forms, as liquids at either a $\times 1$ or $\times 10$ concentration and in powdered form, the latter having the longest shelf life. If the powdered form is to be used it must be reconstituted in water suitable for tissue culture, that is, sterile, double-distilled and deionized.

Since human and animal cell growth is optimal in the pH range 7.2 to 7.4, to operate effectively the buffering system should have its pK_a close to these values. The pK_a of the most commonly used system, sodium bicarbonate, is 6.3 at 37 °C which means that buffering of the growth medium is suboptimal in the physiological pH range and as the bicarbonate dissociates, CO_2 is released into the atmosphere with a resultant increase in alkalinity due to hydroxyl ion formation. To control the release of CO_2 and hence the tendency of the medium pH to rise, the atmosphere of the culture flasks is flushed with a 5% CO_2/95% air mixture prior to commencing incubation or in the case of a CO_2 incubator, throughout the growth period. Alternatively a buffering system based on HEPES (*N*-2-hydroxyethyl-piperazine-*N*^{1}-2-ethanesulphonic acid) with a pK_a at 37 °C of 7.3 has been used satisfactorily as a buffer in the growth of many cell lines; however, it cannot totally replace bicarbonate which besides its buffering role also acts as an essential growth factor for most mammalian cells. In addition HEPES has the disadvantage of causing a granularity of unknown nature in the cytoplasm of certain diploid lines. Therefore, despite its imperfections sodium bicarbonate in conjunction with a CO_2-enriched atmosphere is still the most widely applicable buffering system.

Before commencing culture the antibiotics penicillin and streptomycin are generally added to suppress bacterial contaminants (see Section 7.3.8). Antifungal antibiotics such as nystatin and amphotericin B may also be included but they can be cytotoxic and are best retained for use only if an irreplacable culture is contaminated. The amino acid glutamine should also be added immediately prior to use as it is notably heat-labile, decomposing rapidly at temperatures above −20 °C.

The final ingredient which is essential for the growth of all diploid cell lines is serum, 20% v/v for the establishment of a primary culture and thereafter used at 10% v/v in routine culture. A wide variety of human, animal and avian sera have been employed in routine cell culture the most popular being fetal bovine serum (FBS), although it is a matter of some conjecture as to whether it really is the best source of serum supplement available in terms of growth stimulation or a prime example of user inertia. The exact nature of the role of the serum has been the subject of numerous research papers and reviews, often contradictory and only occasionally enlightening. One critical step in which the serum is involved is ensuring the adequate adhesion of diploid cells to the surface of the support medium. In the absence of protein and divalent cations cell attachment appears to occur by non-specific adsorption alone (Grinnell *et al.*, 1977). The serum supplement provides glycoproteins which actively promote cell attachment and spreading in a manner analogous to that of fibronectin, a protein secreted by many cell lines (Grinnell and Feld, 1979) and required for the development of focal adhesion sites (Virtanen *et al.*, 1982). The serum additionally may supply a wide range of supplementary micronutrients beneficial to cell maintenance and growth but marked differences in the biochemical composition of various batches of serum even within the same species (Bittles, 1974) make this difficult to establish. In view of this observation and the discrepancies in plating efficiency often seen between different serum batches, it is advisable to pretest a range of serum samples derived from alternative pools of the same source-species before placing a bulk order. Obviously if a sequential series of biochemical experiments are to be performed on a cell line, the same batch of serum must be used throughout.

More highly defined media have been compounded, for example, containing nucleic acids, citric acid cycle intermediates and non-essential amino acids, with the aim of improving growth and/or reducing the serum requirement. Supplements such as tryptose phosphate broth also have been used in the culture of specific cell lines and with certain basic media. For the routine subcultivation of most diploid and transformed cultures basic standard growth media, Dulbecco's or the Glasgow modifications of Eagle's medium supplemented with 10% v/v fetal bovine serum, provide satisfactory results. Only when dealing with low cell numbers or tissue from an older donor should it be necessary to utilize one of the more highly defined and expensive preparations, examples being Medium 199 and Ham's F10.

7.3.5 Subcultivation and Harvesting

Under normal growth conditions diploid cells reach confluence after a 1 in 3 subcultivation in 5–7 days and during this period it is usual practice to change the growth medium by pouring and/or pipetting on days 3 or 4 so that essential factors in the medium are not depleted to growth-limiting levels. To minimize cold shock to the cells the replacement medium should always be prewarmed to 37 °C before adding to the culture. It must be emphasized that in any cell culture manipulation, asepsis is more important than absolute volumetric accuracy and pipetting, which should always be carried out using sterile, plugged pipettes, is no exception to this rule. Disposable glass or polystyrene pipettes are more convenient in small- or medium-scale culture facilities; with their greater rigidity, glass pipettes are easier to handle and after sterilization and washing can be re-employed in the general laboratory.

Once the cells reach confluence a subcultivation is necessary. It can be carried out either mechanically by gently detaching the cells from the support medium with a silicone 'policeman' (a silicone bung on the end of a plain glass rod), enzymatically with trypsin or chemically by ethylenediamine tetra-acetate (EDTA) chelation of the divalent cations involved in cell adhesion. The method of cell detachment chosen is dependent on the use to which the cells or their products, including viruses, will be put and it is best primarily to consider this factor as the effect on yield may be considerable. Each will cause some degree of trauma to the cells but for general subcultivation, washing with phosphate-buffered saline Ca^{2+}, Mg^{2+}-free, pH 7.4 at 37 °C to remove all traces of the medium followed by treatment with a combination of trypsin 0.5 gl^{-1} with EDTA 0.2 gl^{-1} is rapid in its action, causes minimal cell loss, and can be used when harvesting cells for many intracellular enzyme assays. If collecting cells for the analysis of surface components proteolytic treatment with trypsin and chelation would not be appropriate and so scraping is the method of choice. In fact scraping also has been found to give much higher yields of cell contents, for example, amino acids from the intracellular pool, than does the trypsin-EDTA treatment (Sambuy, 1982).

When trypsinization is used it should be under optimal conditions, that is at 37 °C and between pH 7.0 and 8.0. In static and roller cultures the cells normally detach within 2–3 minutes but longer periods may be required for microcarriers with a consequent risk to the integrity of the cells, especially affecting their surface components as noted above. Before their introduction to routine use all new batches of trypsin should be tested both for cytotoxicity and proteolytic efficiency as large between-batch variations are quite common. Since solutions of the enzyme can undergo self-digestion they should be stored in small aliquots at −20 °C for no longer than one year.

7.3.6 Post-harvest Cell Rupture

The two main methods of rupturing cultured cells are sonication and homogenization. Although sonication can give rise to marked local heating effects, if carried out at maximum amplitude in 5 to 10 second bursts at 0 °C the damage to cellular constituents is small. The procedure is more rapid and the results are of a greater order of reproducibility than those obtained with homogenization.

7.3.7 Subzero Cell Storage

Diploid and heteroploid cell lines can be stored at low temperatures for many years without detectable changes in their chromosomal structure or in their growth characteristics. The major problems to be avoided in reducing the temperature of the cells to between −130 °C and −196 °C arise mainly from intracellular ice formation damaging cell organelles (Meryman, 1963, 1974; see also Chapter 5) and more particularly the increased concentration of electrolytes within the cells (Lovelock, 1953; Farrant, 1966; Meryman, 1974). To overcome these problems a cryoprotectant, either dimethylsulphoxide (DMSO) or glycerol, is incorporated into the storage medium and the rate of cooling rigidly controlled. For diploid cells an appropriate storage medium is basic growth medium 70%, fetal bovine serum 20%, DMSO 10% plus penicillin and streptomycin, with a cooling rate for a 1 ml vial of 1.5 to 3.0 °C per minute. DMSO is preferred to glycerol because of its more rapid penetration into the cells and greater protective effect.

As might be predicted cell damage can also occur in thawing the cells but it can be reduced to a minimum by as rapid a rate of thawing as possible (McGann and Farrant, 1976). In practice, this is achieved by transferring the vials of cells directly from the nitrogen refrigerator into a 37 °C water bath and then plating their contents in culture medium supplemented with 20% v/v FBS to overcome the slight cytotoxicity of the DMSO. After 6 hours at 37 °C the medium can be replaced with the usual 10% v/v serum-supplemented growth medium and the cells grown to confluence. The recovery rate following subzero storage is usually in the range of 80–95%.

7.3.8 Contamination of Cultures

The growth media are specifically formulated to support the proliferation of diploid and heteroploid cells in culture but they also provide an excellent source of nutrition for the rapid growth of a wide range of micro-organisms including bacteria, fungi, yeasts and *Mycoplasma*. By a judicious mix of asceptic handling techniques and the routine use of penicillin and streptomycin, bacterial contamination is normally a rare occurrence although it should be noted that the incorporation of these antibiotics into the medium can significantly reduce cell yield (Goetz *et al.*, 1979). For this reason and to prevent chronic, low level bacterial contamination affecting their cultures some workers prefer to omit antibiotics from the growth medium.

Mycoplasma infection is a very serious problem in cell culture. Surveys of cell lines from a variety of sources have shown that from 60% (Hayflick, 1965) to 90% (Barile, 1973) may be infected, the highest rates being detected in long-term cultures. Although *Mycoplasma* can alter the metabolism of cultured cells (Stanbridge *et al.*, 1975; States *et al.*, 1978) they may go undetected because the contaminated cultures continue to grow satisfactorily and appear normal under light microscopy. The major sources of *Mycoplasma* infection would appear to be humans, the FBS supplement and possibly the trypsin used in subcultivation which is porcine in origin (Barile *et al.*, 1973). It is strongly recommended that all cultures be tested for the presence of *Mycoplasma* on a regular basis, at least monthly. The fluorescent dye method of Chen (1977) is simple, sensitive, and rapid, showing the *Mycoplasma* as brightly staining particles in the cytoplasm, around the nucleus or in the intercellular spaces either singly, lined in rows, or aggregated in clusters. Tylosin has been recommended as a suitable anti-*Mycoplasma* agent but the only sure way of terminating the infection is by removing and sterilizing all the affected cultures and where feasible this policy should be followed. The hazards in biomedical work from this source of contamination cannot be overestimated as a recent report on the potential effects of *Mycoplasma* in the prenatal diagnosis of argininosuccinic aciduria clearly demonstrated (Fensom *et al.*, 1980).

Contamination may arise from the overgrowth of one cell line by another and in laboratories in which both diploid and faster-growing heteroploid cell lines are cultured it has occurred. The simplest answer to this potential source of contamination is the use of disposable culture vessels and pipettes with the different cell lines being cultured on alternative days. Regular karyotyping and immunological and biochemical analyses should also be used to detect any possible cross-contamination.

Viruses may be detected in cultures, having arisen either as a result of their presence in the original source tissue or as accidental laboratory contaminants. They may exhibit a cytopathogenic effect or may be detectable only after a systematic search involving cocultivation with susceptible cells, RNA–DNA hybridization, and immunological analyses. In any event, cultures suspected of accidently harbouring viruses should immediately be removed and sterilized.

7.4 Applications

The applications of cell culture in biological and biomedical research are widespread as is evidenced by the ever-increasing flow of publications based on the technique in one or other of its many forms. An example of the manner in which the different types of cell culture can inter-relate is provided by studies of the inborn error of amino acid metabolism homocystinuria, due to cystathionine β-synthase deficiency. For heterozygote detection both phytohaemagglutinin-stimulated lymphocytes (Goldstein *et al.*, 1972) and skin fibroblasts (Fleisher *et al.*, 1973; Bittles and Carson, 1981a) have been cultured and the latter also have been employed in investigating the defect at the molecular level (Uhlendorf *et al.*, 1973; Griffiths and Tudball, 1977; Fowler *et al.*, 1978; Bittles and Carson, 1981b). Amnion fibroblasts are used for prenatal diagnosis (Bittles and Carson, 1973; Fleisher *et al.*, 1974) and with the same purpose in mind a technique for the culture of lymphocytes obtained by fetal blood sampling has recently been developed (Fensom, personal communication). In the near future it is probable that opportunities for studying the disorder will be further extended by culturing explants of large organ tissue obtained from fetal as well as post-natal sources, hepatocyte cultures being of particular interest and relevance in this respect.

Further references to detailed culture techniques in current use are provided in Appendix 2 together with a listing of the principal specialist journals in the field.

7.5 Advantages and Disadvantages

Cell culture methods offer the investigator the advantages of a versatile, specific and relatively inexpensive system capable of providing rapid results on material derived from tiny explants of donor tissue. The tissue which can be taken from human, animal or avian sources by techniques of low invasiveness can be cultured for extended periods in the laboratory to produce very large numbers of cells and/or stored over many years at low temperature without detectable change. With cell culture it is possible to study human disease processes using material of human origin rather than via animal models. Given the current climate of opinion on animal experimentation, its avoidance is a useful secondary benefit.

In terms of possible disadvantages, the diploid cell systems are based on monolayer cultures dependent on artificial growth media and removed from the homeostatic control to which they would be subject *in vivo*. All results obtained from *in vitro* experiments must be assessed in this light.

Appendix 1

Glossary of Terms Used

Based on the recommendations of the Committee on Terminology of the Tissue Culture Association of the United States in their most recent, revised form (Schaeffer, 1979).

Anchorage-dependent cells or cultures Cells or cultures derived from them, which will grow, survive or maintain function only when attached to an inert surface such as glass or plastic.

Cell culture This term is used to denote the growing of cells *in vitro*, including the culture of single cells. In cell cultures the cells are no longer organized into tissues.

Cell line A cell line arises from a primary culture at the time of the first subculture. The term cell line implies that cultures from it consist of numerous lineages of cells originally present in the primary culture.

Cell strain A cell strain is derived either from a primary culture or a cell line by the selection or cloning of cells having specific properties or markers. The properties and markers must persist during subsequent cultivation.

Contact inhibition See 'density-dependent inhibition of growth'.

Density-dependent inhibition of growth Mitotic inhibition correlated with the increased cell density.

Diploid The state of the cell in which all chromosomes, except sex chromosomes, are two in number and structurally identical with those of the species from which the culture was derived.

Explant Tissue taken from its original site and transferred to an artificial medium for growth.

Heteroploid The term given to a cell culture when the cells comprising the culture possess nuclei containing chromosome numbers other than the diploid number.

Passage The transfer or transplantation of cells from one culture vessel to another. This term is synonymous with the term 'subculture'.

Plating efficiency The percentage of inoculated cells which give rise to colonies when seeded into culture vessels.

Population doubling level The total number of population doublings for a cell line or strain since its initiation *in vitro*.

Primary culture A culture started from cells, tissues or organs taken directly from organisms. A primary culture may be regarded as such until it is subcultured for the first time. It then becomes a 'cell line'.

Subculture See 'passage'.

Tissue culture The maintenance or growth of tissues *in vitro*, in a way that may allow differentiation and preservation of the architecture and/or function.

Appendix 2

Further Reading

Technical Details and Recent Advances

Cell and Tissue Culture (1975), 5th edn. J. Paul. Churchill Livingstone, Edinburgh

Methods in Cell Biology (1980) Volume 21: Normal Human Tissue and Cell Culture. A Respiratory, Cardiovascular and Integumentary Systems. B Endocrine, Urogenital and Gastrointestinal Systems. eds C. C. Harris, B. F. Trump and G. D. Stoner. Academic Press, New York and London

Microcarrier Cell Culture, Principles and Methods (1981) Pharmacia Fine Chemicals, Uppsala

Specialist Journals on Cell Culture

Cell
Cell and Tissue Research
Experimental Cell Research
In Vitro
Journal of Cell Biology
Journal of Cell Physiology
Journal of Cell Science
Tissue and Cell

Appendix 3

Commercial Sources of Equipment and Supplies

Laminar Flow Cabinets

Hepaire Manufacturing Ltd, 48/50 Fowler Road, Hainault, Ilford, Essex, 1G6 3XA, UK

Microflow Pathfinder Ltd, Fleet Mill, Minley Road, Fleet, Hampshire, GU13 8RD, UK

Incubators

A. Gallenkamp and Co. Ltd, P O Box 290, Technico House, Christopher Street, London, EC2P 2ER, UK

Microscopes

A. Gallenkamp and Co. Ltd, P O Box 290, Technico House, Christopher Street, London, EC2P 2ER, UK

Liquid Nitrogen Refrigerators

Union Carbide UK Ltd,
Distributors: Jencons Scientific Ltd, Cherrycourt Way Industrial Estate Stanbridge Road, Leighton Buzzard, Beds, LU7 8UA, UK

Pipettes

Sterilin Ltd, 43–45 Broad Street, Teddington, Middlesex, TW11 8QZ, UK

Volac, John Poulten Ltd, Tanner Street, Barking, Essex, IG11 8QD, UK

Culture Vessels and Growth Media

Flow Laboratories Ltd, P O Box 17, Second Avenue Industrial Estate, Irvine, Ayrshire, Scotland, KA12 8NB, UK
and
7655 Old Springhouse Road, McLean, Virginia 22102, USA

Gibco Europe Ltd, P O Box 35, Washington Road, Abbotsinch Industrial Estate, Paisley PA3 4EP, Scotland, UK

Chemicals

BDH Chemicals Ltd, Baird Road, Enfield, Middlesex EN1 1SH, UK

Hopkins and Williams Freshwater Road, Chadwell Heath, Essex, UK

References

Abercrombie, M. (1979) Contact inhibition and malignancy. *Nature* **281**, 259–262

Barile, M. F. (1973) In: *Contamination of Cell Cultures* (ed. J. Fogh), pp. 131–172. Academic Press, New York

Barile, M. F., del Giudice, R. A., Hopps, H. E., Grabowsky, M. W., and Riggs, D. B. (1973) The identification and sources of *mycoplasmas* isolated from contaminated cell cultures. *Ann. N.Y. Acad. Sci.* **225**, 251–264

Bittles, A. H. (1974) The comparative analysis of three batches of foetal bovine serum used in tissue culture. *Med. Lab. Technol.* **31**, 253–255

Bittles, A. H., and Carson, N. A. J. (1973) Tissue culture techniques as an aid to prenatal diagnosis and genetic counselling in homocystinuria. *J. Med. Genet.* **10**, 120–121

Bittles, A. H., and Carson, N. A. J. (1981a) Homocystinuria: studies on cystathionine β-synthase, *S*-adenosylmethionine synthetase and cystathionase activities in skin fibroblasts. *J. Inherit. Metab. Dis.* **4**, 3–6

Bittles, A. H., and Carson, N. A. J. (1981b) Homocystinuria: the effect of pyridoxine supplementation on cultured skin fibroblasts. *J. Inherit. Metab. Dis.* **4**, 7–9

Borysenko, J. Z., and Woods, W. (1979) Density, distribution and mobility of surface anions on a normal/transformed cell pair. *Exp. Cell Res.* **118**, 215–227

Burrows, M. T. (1910) The cultivation of tissues of the chick-embryo outside the body. *J. Am. Med. Ass.* **55**, 2057–2058

Carrel, A. (1912a) On the permanent life of tissues outside of the organism. *J. Exp. Med.* **15**, 516–528

Carrel, A. (1912b) Pure cultures of cells. *J. Exp. Med.* **16**, 165–168

Chen, T. R. (1977) *In situ* detection of *mycoplasma* contamination in cell cultures by fluorescent Hoechst 33258 stain. *Exp. Cell Res.* **104**, 255–262

Eagle, H. (1955) The specific aminoacid requirements of a mammalian cell (Strain L) in tissue culture. *J. Biol. Chem.* **214**, 839–852

Eagle, H. (1956) The salt requirements of mammalian cells in tissue culture. *Arch. Biochem.* **61**, 356–366

Eagle, H., Barban, S., Levy, M., and Schulze, H. O. (1958) The utilization of carbohydrates by human cell cultures. *J. Biol. Chem.* **233**, 551–558

Eagle, H., and Piez, K. A. (1960) The utilization of proteins by cultured human cells. *J. Biol. Chem.* **235**, 1095–1097

Edwards, J. H. (1960) Painless skin biopsy. *Lancet* **i**, 496

Evans, V. J., Earle, W. R., Sanford, K. K., Shannon, J. E., and Waltz, H. K. (1951) The preparation and handling of replicate tissue cultures for quantitative studies. *J. Natl Cancer Inst.* **11**, 907–912

Farrant, J. (1966) The preservation of living cells, tissues and organs at low temperatures: some underlying principles. *Lab. Pract.* **15**, 402–404, 409

Fensom, A. H., Benson, P. F., Baker, J. E., and Mutton, D. E. (1980) Prenatal diagnosis of argininosuccinic aciduria: effect of *mycoplasma* contamination on the indirect assay for argininosuccinate lyase. *Am. J. Hum. Genet.* **32**, 761–763

Fleisher, L. D., Longhi, R. C., Tallan, H. H., Beratis, N. G., Hirschhorn, K., and Gaull, G. E. (1974) Homocystinuria: investigations of cystathionine synthase in cultured fetal cells and the prenatal determination of genetic status. *J. Pediatr.* **85**, 677–680

Fleisher, L. D., Tallan, H. H., Beratis, N. G., Hirschhorn, K., and Gaull, G. E. (1973) Cystathionine synthase deficiency: heterozygote detection using cultured skin fibroblasts. *Biochem. Biophys. Res. Commun.* **55**, 38–44

Fowler, B., Kraus, J., Packman, S., and Rosenberg, L. E. (1978) Homocystinuria: evidence for three distinct classes of cystathionine β-synthase mutants in cultured fibroblasts. *J. Clin. Invest.* **61**, 645–653

Goetz, I. E., Moklebust, R., and Warren, C. J. (1979) Effects of some antibiotics on the growth of human diploid skin fibroblasts in cell culture. *In Vitro* **15**, 114–119

Goldstein, J. L., Campbell, B. K., and Gartler, S. M. (1972) Cystathionine synthase activity in human lymphocytes: induction by phytohemagglutinin. *J. Clin. Invest.* **51**, 1034–1037

Goldstein, S. (1971) The biology of aging. *New Engl. J. Med.* **285**, 1120–1129

Griffiths, R., and Tudball, N. (1977) The molecular defect in a case of (cystathionine β-synthase-deficient) homocystinuria. *Eur. J. Biochem.* **74**, 269–273

Grinnell, F., and Feld, M. K. (1979) Initial adhesion of human fibroblasts in serum-free medium: possible role of secreted fibronectin. *Cell* **17**, 117–129

Grinnell, F., Hays, D. G., and Minter, D. (1977) Cell adhesion and spreading factor. Partial purification and properties. *Exp. Cell Res.* **110**, 175–190

Hayflick, L. (1965) The limited *in vitro* lifetime of human diploid cell strains. *Exp. Cell Res.* **37**, 614–636

Hayflick, L., and Moorhead, P. S. (1961) The serial cultivation of human diploid cell strains. *Exp. Cell Res.* **25**, 585–621

Lovelock, J. E. (1953) The mechanism of the protective action of glycerol against haemolysis by freezing and thawing. *Biochim. Biophys. Acta* **11**, 28–36

McGann, L. E., and Farrant, J. (1976) Survival of tissue culture cells frozen by a 2-step procedure to −196 °C. *Cryobiol.* **13**, 269–273

Maroudas, N. G. (1975) Adhesion and spreading of cells on charged surfaces. *J. Theor. Biol.* **49**, 417–424

Martin, G., Sprague, C., and Epstein, C. (1970) Replicative life-span of cultivated human cells. Effects of donor's age, tissue and genotype. *Lab. Invest.* **23**, 86–92

Meryman, H. T. (1963) Preservation of living cells. *Fed. Proc.* **22**, 81–89

Meryman, H. T. (1974) Freezing injury and its prevention in living cells. *Ann. Rev. Biophys. Bioeng.* **3**, 341–363

Roux, W. (1885) Beiträge der entwicklungsmechanik des embryo. *Z. Biol.* **21**, 411–526

Sambuy, Y. (1982) The effects of ageing processes on the free aminoacid pool of cultured human diploid fibroblasts. Ph.D. thesis, University of London

Sanford, K. K., Earle, W. R., Evans, V. J., Waltz, H. K., and Shannon, J. E. (1951) The measurement of proliferation in tissue cultures by enumeration of cell nuclei. *J. Natl Cancer Inst.* **11**, 773–795

Schaeffer, W. I. (1979) Proposed usage of animal tissue culture terms (revised 1978). Usage of vertebrate cell, tissue and organ culture terminology. *In Vitro* **15**, 649–643

Stanbridge, E. J., Tischfield, J. A., and Schneider, E. L. (1975) Appearance of hypoxanthine guanine phosphoribosyl-transferase activity as a consequence of *mycoplasma* contamination. *Nature* **256**, 329–331

States, B., Harris, D., Hummeler, K., and Segal, S. (1978) Aminoacid disorders in tissue culture cells: effects of *mycoplasma* and antibiotics on cystine metabolism. *Monogr. Hum. Genet.* **9**, 131–134

Uhlendorf, B. W., Conerly, E. B., and Mudd, S. H. (1973) Homocystinuria: studies in tissue culture. *Pediatr. Res.* **7**, 645–658

van Wezel, A. L. (1967) Growth of cell-strains and primary cells on micro-carriers in homogeneous culture. *Nature* **216**, 64–65

Virtanen, I., Vartio, T., Badley, R. A., and Lehto, V.-P. (1982) Fibronectin in adhesion, spreading and cytoskeletal organisation. *Nature* **298**, 660–663

Index